Collins
Revision

NEW GCSE

Physics

Foundation and Higher

for Edexcel

Authors: **Sarah Mansel**
Caroline Reynolds

Revision Guide +
Exam Practice Workbook

Contents

Contents

The Solar System

Models of the Solar System

G–E

- The ancient Greeks used a **geocentric model** of the **Solar System**, proposed by Ptolemy.
- In the geocentric model, the Earth was in the centre and everything else (planets, the Moon, the Sun and the stars) moved around the Earth in circular orbits.
- Ptolemy explained the occasional **retrograde motion** of Mars using epicycles.
- Ptolemy's model of the Solar System was very complicated and did not accurately predict the position of the planets.

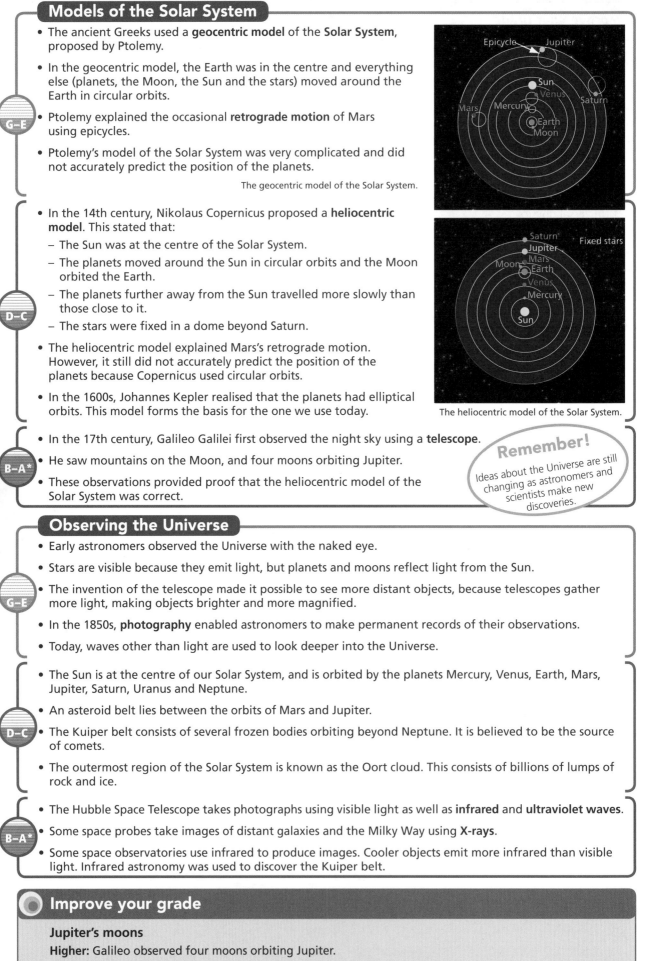

The geocentric model of the Solar System.

D–C

- In the 14th century, Nikolaus Copernicus proposed a **heliocentric model**. This stated that:
  - The Sun was at the centre of the Solar System.
  - The planets moved around the Sun in circular orbits and the Moon orbited the Earth.
  - The planets further away from the Sun travelled more slowly than those close to it.
  - The stars were fixed in a dome beyond Saturn.
- The heliocentric model explained Mars's retrograde motion. However, it still did not accurately predict the position of the planets because Copernicus used circular orbits.
- In the 1600s, Johannes Kepler realised that the planets had elliptical orbits. This model forms the basis for the one we use today.

The heliocentric model of the Solar System.

B–A*

- In the 17th century, Galileo Galilei first observed the night sky using a **telescope**.
- He saw mountains on the Moon, and four moons orbiting Jupiter.
- These observations provided proof that the heliocentric model of the Solar System was correct.

> **Remember!**
> Ideas about the Universe are still changing as astronomers and scientists make new discoveries.

Observing the Universe

G–E

- Early astronomers observed the Universe with the naked eye.
- Stars are visible because they emit light, but planets and moons reflect light from the Sun.
- The invention of the telescope made it possible to see more distant objects, because telescopes gather more light, making objects brighter and more magnified.
- In the 1850s, **photography** enabled astronomers to make permanent records of their observations.
- Today, waves other than light are used to look deeper into the Universe.

D–C

- The Sun is at the centre of our Solar System, and is orbited by the planets Mercury, Venus, Earth, Mars, Jupiter, Saturn, Uranus and Neptune.
- An asteroid belt lies between the orbits of Mars and Jupiter.
- The Kuiper belt consists of several frozen bodies orbiting beyond Neptune. It is believed to be the source of comets.
- The outermost region of the Solar System is known as the Oort cloud. This consists of billions of lumps of rock and ice.

B–A*

- The Hubble Space Telescope takes photographs using visible light as well as **infrared** and **ultraviolet waves**.
- Some space probes take images of distant galaxies and the Milky Way using **X-rays**.
- Some space observatories use infrared to produce images. Cooler objects emit more infrared than visible light. Infrared astronomy was used to discover the Kuiper belt.

Improve your grade

Jupiter's moons

Higher: Galileo observed four moons orbiting Jupiter.

Explain how this observation did not fit with the geocentric model of the Solar System.   *AO2* [3 marks]

Reflection, refraction and lenses

Reflection

- All waves can be **reflected**:
  - Sound waves are reflected off walls and buildings (echoes).
  - Dolphins and bats use ultrasound (**echolocation**) to monitor their surroundings.
  - Radio waves are reflected from surfaces within the atmosphere such as the ionosphere.
- Light waves will reflect from shiny surfaces following the laws of reflection:
  - The angle of incidence (*i*) equals the angle of reflection (*r*).
  - The incident ray, the reflected ray and the normal are all in the same plane.

Remember!
Always measure the angle from the normal to the ray.

angle *i* = angle *r*

normal

incident ray   *i*   *r*   reflected ray

plane mirror

A ray diagram showing reflection.

G–E

Refraction

- **Refraction** is the bending of light rays at a surface, for example when entering or leaving a glass block.
- When light travels from air to glass, the direction of the ray bends towards the normal. The angle of incidence is larger than the angle of refraction.
- All waves are refracted at the boundary between different materials.

normal

weak reflection   incident ray

air   *i*

glass   *r*

refracted ray

A ray of light is refracted towards the normal when it travels from air into glass.

D–C

- Refraction occurs because the **speed** of the wave changes as it passes through different materials.
- The speed of light depends on the **density** of the material. The speed of light in air is almost 3.0×10^8 m/s, but in water it is closer to 2.0×10^8 m/s.
- When light travels from a less dense material to a more dense material it gets slower. This causes it to bend towards the normal.
- When light travels from a more dense material to a less dense material it speeds up. This causes it to bend away from the normal.

B–A*

Understanding lenses

- There are two types of lens:
  - A **converging** (or convex) **lens** is fatter in the middle.
  - A **diverging** (or concave) **lens** is thinner in the middle.
- Converging lenses are used to form images in telescopes, cameras, projectors, binoculars and our eyes.
- Rays of light that are parallel to the principal axis of a converging lens are refracted inwards. They converge on the **principal focus** (focal point, F).
- The distance between the centre of the lens and the focal point is called the **focal length** (*f*) of the lens.
- The fatter the lens, the shorter the focal length.

converging lens

diverging lens

principal focus (focal point)

principal axis

F

focal plane

focal length (*f*)

G–E

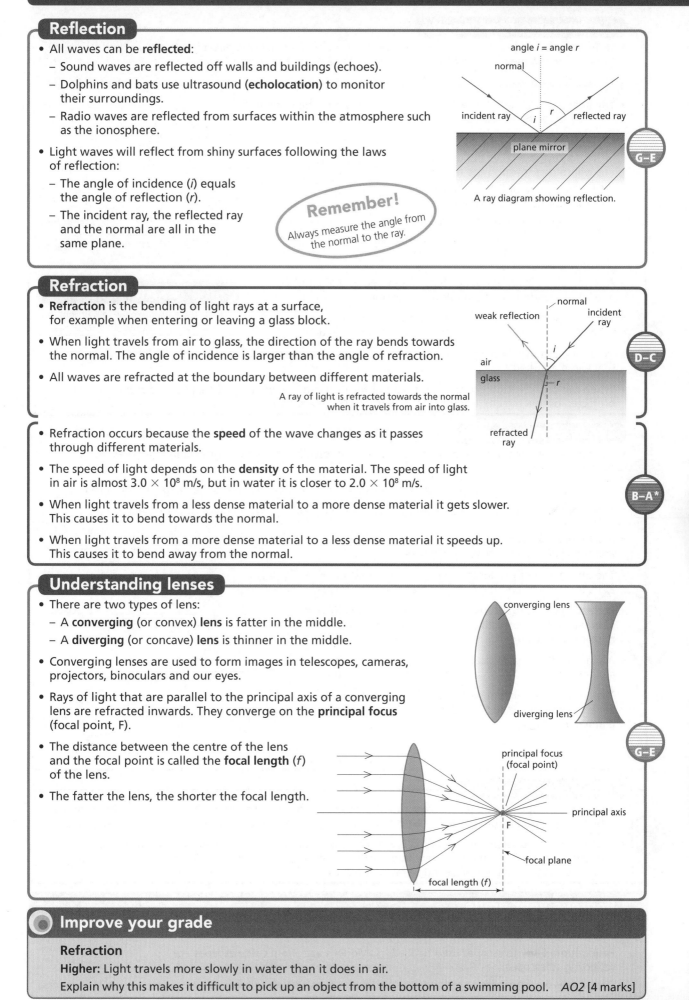

Improve your grade

Refraction

Higher: Light travels more slowly in water than it does in air.

Explain why this makes it difficult to pick up an object from the bottom of a swimming pool.   *AO2* [4 marks]

Lenses in telescopes

Determining focal length

- To find the focal length of a converging lens you obtain an image of a distant object (tree, building, etc.) on a screen and measure the distance between the lens and the focused image.
- This works because the rays of light from a distant object are parallel and will converge at the focal point.
- The image you obtain is inverted (upside down), **diminished** (smaller than the object) and **real** (can be projected onto a screen).
- On ray diagrams:
  - Draw a ray parallel to the principal axis. It will refract through the focal point.
  - Draw a ray straight through the centre of the lens (it is not deviated).
  - The image is formed where the two rays cross.

A ray diagram showing how to find the properties of an image.

- A converging lens can produce both **magnified** and diminished images, depending on the position of the object.
- When the distance between the object and the lens is greater than 2f the image will be inverted, real and diminished.
- When the distance between the object and the lens is between f and 2f, the image will be inverted, real and magnified.
- When the object is closer to the lens than the focal length, the image is **virtual** (it cannot be projected onto a screen), upright and magnified. This is a magnifying glass.

EXAM TIP
Always use a ruler to draw rays of light – they travel in straight lines.

Types of telescopes

- Early telescopes had a converging lens at the front (**objective lens**) and a diverging lens (**eyepiece**) to look through.
- Lenses have different focal points for different colours of light. This makes the image a bit blurred.
- A clearer image can be formed by using a concave (parabolic) mirror in place of the objective lens.

- In a modern **refracting telescope**, both the objective lens and the eyepiece are converging lenses.
- The objective lens produces an image of a distant object at its focal point, and the eyepiece magnifies the image.
- **Reflecting telescopes** can be much larger and are easier to manoeuvre than refracting telescopes. This is because a mirror has much smaller mass than a bulky lens.
- A simple reflecting telescope uses a large concave mirror, a plane mirror and a converging lens.
- The concave mirror forms an image of a distant object, which is then reflected towards the eyepiece using the plane mirror.

A refracting telescope.

A reflecting telescope.

Telescopes in space

- Modern telescopes are huge. They are usually housed in observatories on high mountains, where there is little light pollution from cities. The air is also cooler, so there is less interference from the atmosphere.
- The Hubble space telescope is orbiting Earth. There is no atmospheric interference in space, and the telescope can produce images of very faint distant objects up to magnifications of about 5000. It also produces images using infrared and ultraviolet waves.

Improve your grade

Advantages of modern telescopes
Foundation: Early telescopes used two lenses. Modern astronomical telescopes are reflecting telescopes.
Describe **two** advantages of reflecting telescopes over refracting telescopes.

AO1 [4 marks]

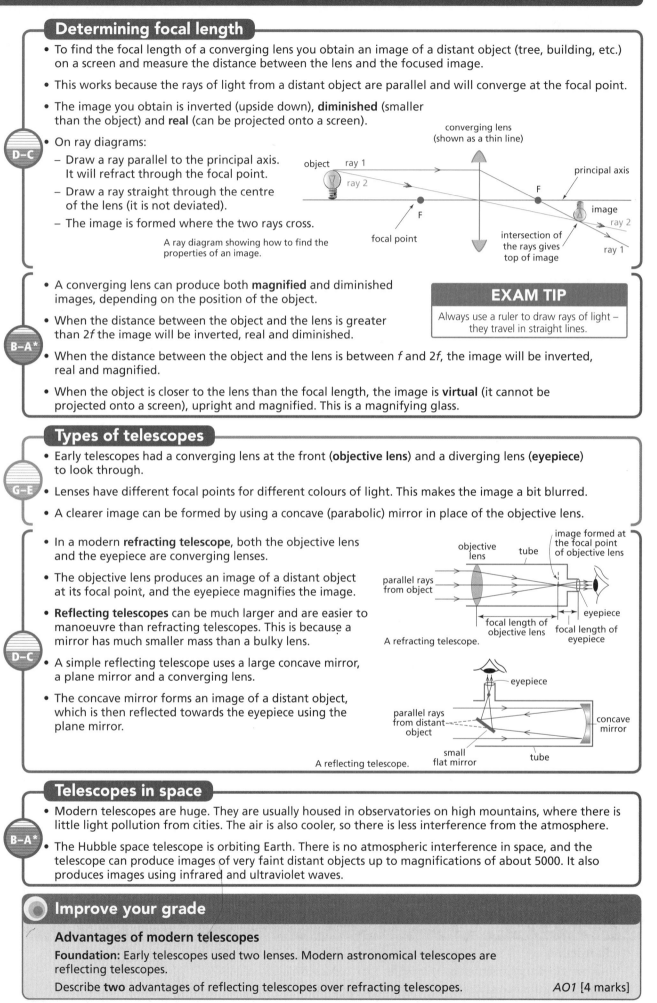

Waves

Understanding waves

- Waves transfer energy and information from one place to another using vibrations, but they do not transfer matter in the direction they are travelling.

- The **wavelength** (in metres) of a wave is the distance the wave travels in one complete cycle – the distance between two adjacent **peaks** or two adjacent **troughs**.

- The **amplitude** (in metres) of a wave is the maximum **displacement** of a wave. It is measured from the top of a peak to the centre line, or from the bottom of a crest to the centre line.

- The **frequency** (in hertz) of a wave is the number of complete waves passing a point in one second.

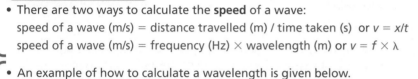

Important things you need to know about a wave.

G–E

Wave equations

- There are two ways to calculate the **speed** of a wave:

  speed of a wave (m/s) = distance travelled (m) / time taken (s) or $v = x/t$

  speed of a wave (m/s) = frequency (Hz) × wavelength (m) or $v = f \times \lambda$

D–C

- An example of how to calculate a wavelength is given below.
  - Sound waves travel through air at a speed of 340 m/s. A particular note of sound has a frequency of 260 Hz. To calculate the wavelength:

    speed = frequency × wavelength

    wavelength = speed / frequency

    wavelength = 340 / 260 = 1.3 m

B–A*

Types of waves

- All waves belong to one of two groups:
  - In **transverse waves**, the vibration is at right angles to the direction of wave travel.
  - In **longitudinal waves**, the vibration is back and forth along the direction of wave travel.

- Electromagnetic waves and water waves are transverse waves.

- Sound waves are longitudinal waves.

transverse wave

longitudinal wave

compressions

You can use a Slinky to create both transverse and longitudinal waves.

G–E

Seismic waves

- Longitudinal waves such as sound waves travel as a series of **compressions** (areas of higher pressure) and **rarefactions** (areas of lower pressure).

- **Seismic waves** are produced by earthquakes and explosions. There are two main types:
  - Primary (P) waves are slower-moving longitudinal waves.
  - Secondary (S) waves are faster-moving transverse waves.

> **Remember!**
> The wavelength of a longitudinal wave is the distance from one compression to the next compression.

D–C

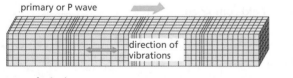

primary or P wave

direction of vibrations

secondary or S wave

direction of vibrations

Two types of seismic waves.

- Seismic waves are detected using a seismograph or **seismometer**.

- The vibration of the Earth's surface is recorded by its motion relative to a heavy pendulum.

B–A*

Improve your grade

Thunder and lightning

Higher: During a thunderstorm you always see the lightning before you hear the thunder. Light travels so fast that the lightning is almost instantaneous, but sound travels at a speed of 340 m/s in air.
If you hear thunder half a minute after seeing lightning, how far away is the storm? *AO3* [3 marks]

The electromagnetic spectrum

Visible light

G–E

- White light is made up of different colours. These colours can be dispersed into a spectrum using a **prism**.

- The colours in the visible spectrum are red, orange, yellow, green, blue, indigo and violet.

- Each colour of light has a different wavelength. Red light has a longer wavelength than violet light.

The visible spectrum.

Infrared and ultraviolet

D–C

- William Herschel discovered **infrared** waves in 1800, while investigating the temperature of the visible spectrum.

- He found that the hottest temperatures were beyond the red end of the spectrum, where there is no visible light. This is known as the infrared region.

- Infrared waves have a longer wavelength than red light.

- Johann Ritter discovered **ultraviolet** waves in 1801, while experimenting with silver chloride used in photography. The rate of reaction was highest beyond the violet end of the visible spectrum.

- Ultraviolet waves have a shorter wavelength than violet light.

B–A*

- Ultraviolet waves from the Sun are harmful. They will damage skin cells and eyes. The most common form of this damage is sunburn, but long exposure can cause skin cancer.

- There are three types of ultraviolet waves, known as UV-A, UV-B and UV-C. UV-C waves have the shortest wavelength and cause the most damage.

EM waves

G–E

- Ultraviolet, visible light and infrared are all part of the family of waves known as the **electromagnetic spectrum**.

- All electromagnetic waves are **transverse** and travel at the same speed in a **vacuum** (300 000 000 m/s).

- As with all other waves, EM waves:
  - can transfer energy
  - can be reflected, refracted and **diffracted**
  - obey the wave equation: wave speed = wavelength × frequency.

D–C

- All EM waves are essentially the same, but they have different wavelengths and frequencies.

- In order of decreasing wavelength and increasing frequency, the EM waves are: radio waves, microwaves, infrared, visible light, ultraviolet, X-rays and gamma rays.

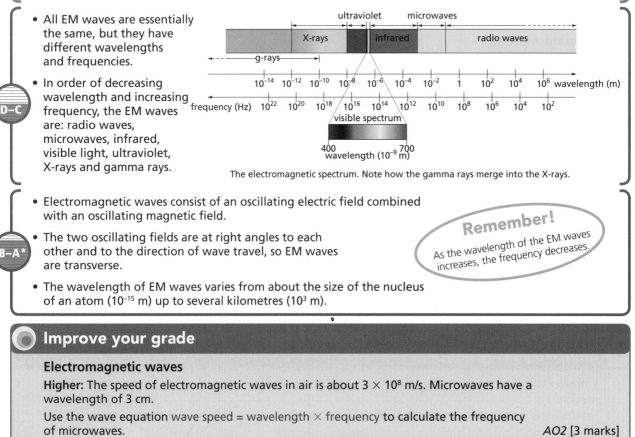

The electromagnetic spectrum. Note how the gamma rays merge into the X-rays.

B–A*

- Electromagnetic waves consist of an oscillating electric field combined with an oscillating magnetic field.

- The two oscillating fields are at right angles to each other and to the direction of wave travel, so EM waves are transverse.

- The wavelength of EM waves varies from about the size of the nucleus of an atom (10^{-15} m) up to several kilometres (10^3 m).

Remember!
As the wavelength of the EM waves increases, the frequency decreases.

Improve your grade

Electromagnetic waves

Higher: The speed of electromagnetic waves in air is about 3×10^8 m/s. Microwaves have a wavelength of 3 cm.

Use the wave equation wave speed = wavelength × frequency to calculate the frequency of microwaves.

AO2 [3 marks]

Topic 2: 2.1, 2.2, 2.3, 2.4, 2.6c

Uses of EM waves

Radio waves and microwaves

- Radio waves are produced and detected by aerials.

- Radio waves are used to broadcast television and radio programmes. They are also used by emergency services for communication.

- Microwaves are very high frequency (short wavelength) radio waves.

- Microwaves are used to cook food in a microwave oven. The microwaves are absorbed by water and fat in the food, and the energy becomes heat.

- Microwaves are also used for mobile-phone communication.

- Medium-frequency radio waves (MW) can reflect off the **ionosphere** to communicate long distances. These reflected waves are called sky waves.

- High-frequency radio waves (and microwaves) travel in straight lines to satellites.

- Mobile-phone networks use satellites to transmit communications around the world at high speeds.

Remember!
Satellite TV is broadcast using very short wavelength radio waves, so satellite dishes need to be pointed directly towards the satellite.

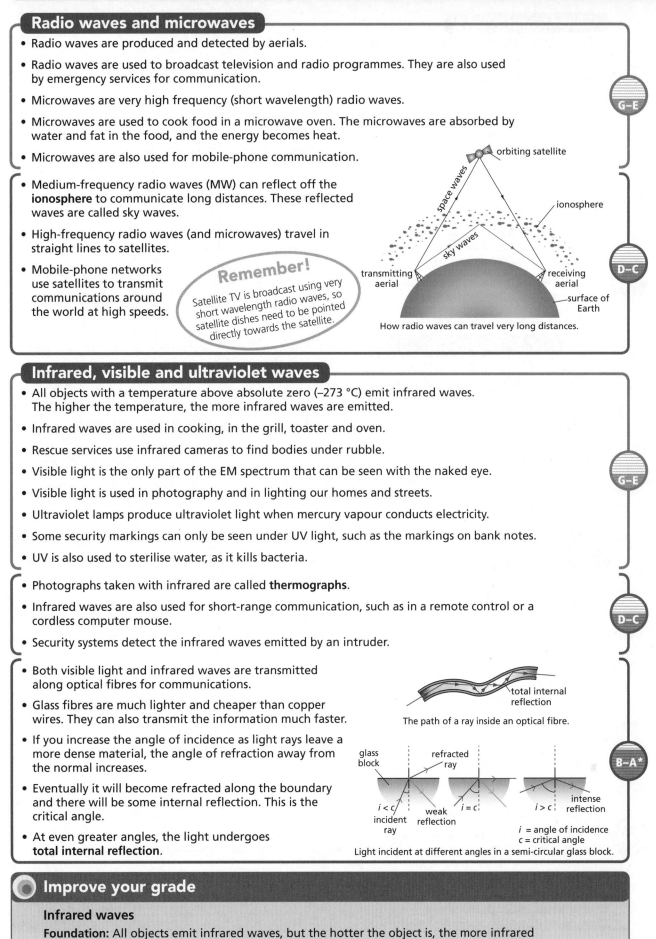

How radio waves can travel very long distances.

Infrared, visible and ultraviolet waves

- All objects with a temperature above absolute zero (−273 °C) emit infrared waves. The higher the temperature, the more infrared waves are emitted.

- Infrared waves are used in cooking, in the grill, toaster and oven.

- Rescue services use infrared cameras to find bodies under rubble.

- Visible light is the only part of the EM spectrum that can be seen with the naked eye.

- Visible light is used in photography and in lighting our homes and streets.

- Ultraviolet lamps produce ultraviolet light when mercury vapour conducts electricity.

- Some security markings can only be seen under UV light, such as the markings on bank notes.

- UV is also used to sterilise water, as it kills bacteria.

- Photographs taken with infrared are called **thermographs**.

- Infrared waves are also used for short-range communication, such as in a remote control or a cordless computer mouse.

- Security systems detect the infrared waves emitted by an intruder.

- Both visible light and infrared waves are transmitted along optical fibres for communications.

- Glass fibres are much lighter and cheaper than copper wires. They can also transmit the information much faster.

- If you increase the angle of incidence as light rays leave a more dense material, the angle of refraction away from the normal increases.

- Eventually it will become refracted along the boundary and there will be some internal reflection. This is the critical angle.

- At even greater angles, the light undergoes **total internal reflection**.

The path of a ray inside an optical fibre.

i = angle of incidence
c = critical angle

Light incident at different angles in a semi-circular glass block.

◉ Improve your grade

Infrared waves

Foundation: All objects emit infrared waves, but the hotter the object is, the more infrared radiation is given off.

Use this information to explain how the police could use infrared radiation to find fugitives.

AO2 [3 marks]

Gamma rays, X-rays, ionising radiation

Gamma and X-rays

G–E

- X-rays are produced when fast-moving electrons hit a metal plate. They can be used to detect an object's internal structure.
- Uses of X-rays include:
  - airport security scanners
  - medical X-rays to detect the condition of bones and teeth
  - to detect unwanted pieces of metal in machinery.

D–C

- Gamma rays are similar to X-rays, but they are produced by radioactive materials.
- Gamma rays kill bacteria, so they are used to sterilise food and medical instruments.
- Radioactive **tracers** emitting gamma rays are used to detect some forms of cancer, and gamma rays are used in **radiotherapy** to treat cancers.
- The higher the frequency of electromagnetic radiation, the higher its energy and the more harmful it is. The table opposite shows the effects of exposure to electromagnetic waves.

Waves	What they do
Radio waves	They are safe because they do not produce ionisation.
Microwaves	Cause internal heating of the body cells.
Infrared	Can cause skin burns.
Visible light	Intense light can cause permanent damage to the retina.
Ultraviolet	Intense ultraviolet light damages skin (surface) cells; can trigger skin cancer; can cause eye conditions such as cataracts; can destroy proteins in the eye lens.
X-rays and gamma rays	Can damage the DNA of cells; can mutate cells; can trigger cancer.

EXAM TIP
When discussing the use of ionising radiation, make sure you understand both the risks and the benefits.

B–A*

- In an X-ray machine, a very high voltage is applied between the electrodes.
- The electrons are accelerated and hit a tungsten target.
- The fast-moving electrons collide with tungsten atoms, causing them to emit X-rays.

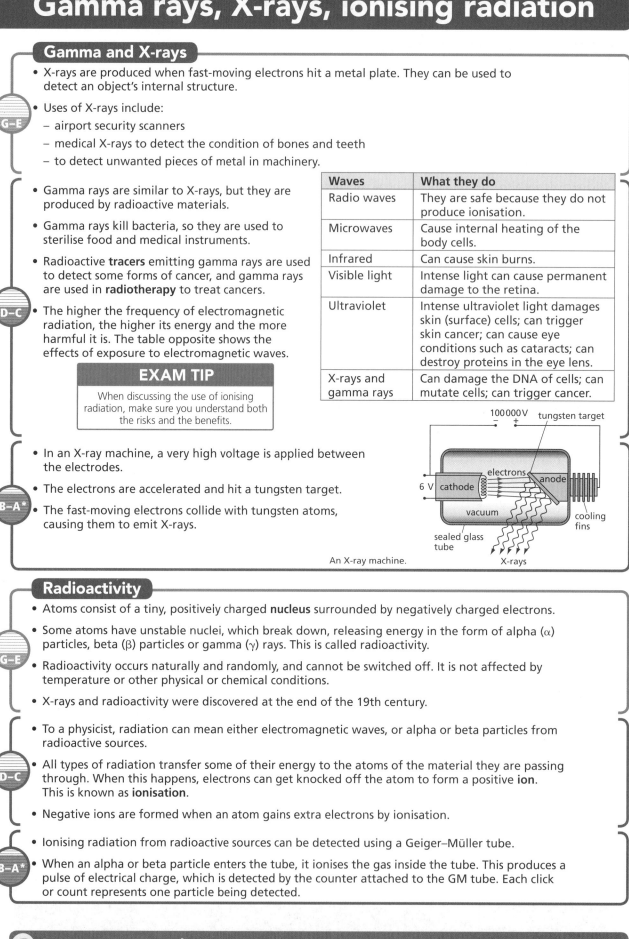

An X-ray machine.

Radioactivity

G–E

- Atoms consist of a tiny, positively charged **nucleus** surrounded by negatively charged electrons.
- Some atoms have unstable nuclei, which break down, releasing energy in the form of alpha (α) particles, beta (β) particles or gamma (γ) rays. This is called radioactivity.
- Radioactivity occurs naturally and randomly, and cannot be switched off. It is not affected by temperature or other physical or chemical conditions.
- X-rays and radioactivity were discovered at the end of the 19th century.

D–C

- To a physicist, radiation can mean either electromagnetic waves, or alpha or beta particles from radioactive sources.
- All types of radiation transfer some of their energy to the atoms of the material they are passing through. When this happens, electrons can get knocked off the atom to form a positive **ion**. This is known as **ionisation**.
- Negative ions are formed when an atom gains extra electrons by ionisation.

B–A*

- Ionising radiation from radioactive sources can be detected using a Geiger–Müller tube.
- When an alpha or beta particle enters the tube, it ionises the gas inside the tube. This produces a pulse of electrical charge, which is detected by the counter attached to the GM tube. Each click or count represents one particle being detected.

Improve your grade

Radioactivity

Foundation: Radioactive materials emit ionising radiation.
Explain what is meant by the term 'ionising radiation'.

AO1 [2 marks]

The Universe

Our place in the Universe

- The **Universe** is made up of billions of **galaxies**, which are collections of stars held together by the force of gravity.

- Our **Solar System** is part of the **Milky Way** galaxy.

- Our Sun is an average star in the Milky Way.

- Distances in the Solar System are measured in Astronomical Units (AU). One AU is the average distance from the Earth to the Sun:

  $1 \text{ AU} = 1.5 \times 10^{11} \text{ m}$

Object	Average distance from the Sun (AU)	Diameter relative to Earth
Sun	–	110
Mercury	0.39	0.38
Venus	0.72	0.95
Earth	1.00	1.00
Moon	1.00	0.27
Mars	1.52	0.53
Jupiter	5.20	11.20
Saturn	9.54	9.44
Uranus	19.20	3.69
Neptune	30.10	3.48

G–E

The scale of things

- The Universe is believed to be about 10^{27} m across.

- It contains about 10^{11} galaxies, which each contain about 10^{11} stars.

- Each star may have its own system of planets, just like our Solar System.

> **Remember!**
> Using standard notation, the power of ten gives the number of zeros you need – for example, 10^3 is 1000 (3 zeros).

Object	Approximate distance (m)	Comments
Earth and Moon	3×10^8	Our nearest neighbour is about 30 Earth diameters away.
Sun and Earth	1.5×10^{11}	It takes about 8 minutes for light from the Sun to reach us.
Solar System	4×10^{12}	The outermost planet Neptune is about 30 times further from the Sun than Earth.
Nearest star	4×10^{16}	Proxima Centauri is the nearest star beyond the Sun. No planets have been detected around it.
Closest galaxy	2×10^{22}	Andromeda is the nearest galaxy and is very similar to the Milky Way. It is about 10^{21} m across.
Size of the Universe	1×10^{27}	The Universe contains all the galaxies in space.

D–C

- Very large distances in space are measured in **light-years**.
  A light-year is the distance travelled by light (through a vacuum) in a year.

- The size of the Universe is about 100 billion light-years.

B–A*

Exploring the Universe

- Modern telescopes use all parts of the electromagnetic spectrum to observe and record the Universe.

- Larger magnification telescopes allow astronomers to view stars and galaxies in deep space.

- Stars and galaxies (some of which cannot be seen using visible light) also emit other types of electromagnetic waves such as X-rays.

G–E

- Space probes are sent out on flyby missions to detect the presence of water and minerals on other planets. They send data and photographs back to Earth using radio waves.

- Robotic **landers** can be sent out to planets. They collect soil samples and search for microscopic life forms, such as bacteria.

- The Search for **Extraterrestrial** Intelligence (SETI) is looking for radio broadcasts from alien civilisations, which may exist on other planets orbiting distant stars.

D–C

- Photographs can be taken using parts of the EM spectrum other than visible light, such as X-rays. These reveal different information about the structure of space bodies, such as their magnetic fields.

- Images like the one of Jupiter opposite are taken with X-rays. They are analysed by computers and given false colours.

Jupiter in X-rays.

B–A*

⬤ Improve your grade

Studying the Universe

Foundation: Humans have always been fascinated with space. Ancient civilisations relied on the naked eye to study the stars. In medieval times, telescopes were used to study objects in space.
Explain **two** advantages of the technology used to study the Universe today.   *AO2* [4 marks]

Analysing light

The spectrometer

- White light can be split up into its component colours by **refraction** using a glass prism. It can also be split up by reflecting it off a shiny surface such as the surface of a DVD.

- A **spectrometer** is a device used for looking at the spectrum of light.

- You can make a simple model spectrometer using a cardboard box and an old DVD, as shown in the diagram opposite.

- To get a clearer image, make the slit narrower and observe the light source in a darkened room.

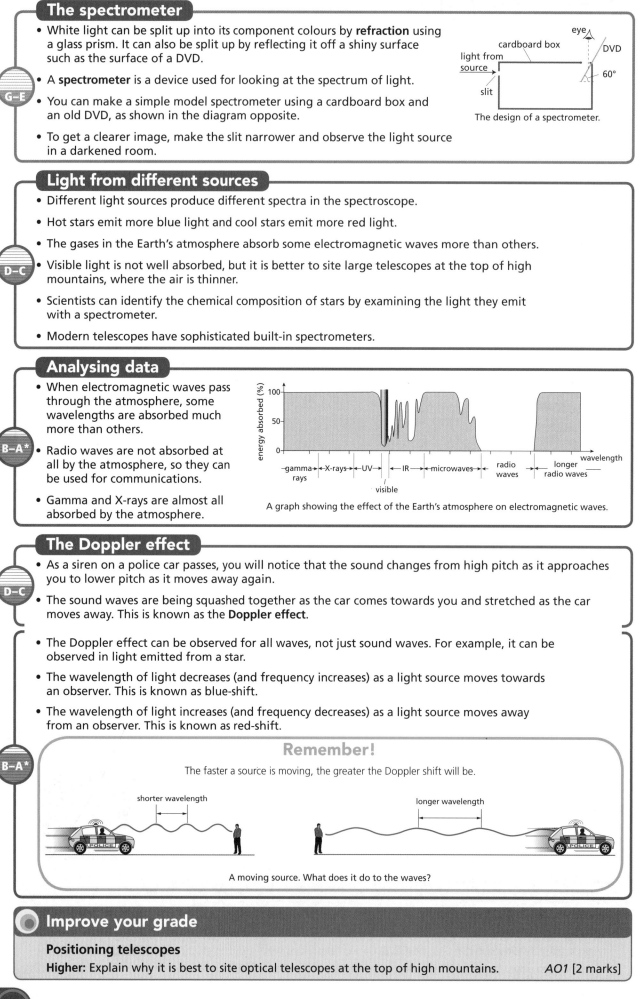

The design of a spectrometer.

Light from different sources

- Different light sources produce different spectra in the spectroscope.

- Hot stars emit more blue light and cool stars emit more red light.

- The gases in the Earth's atmosphere absorb some electromagnetic waves more than others.

- Visible light is not well absorbed, but it is better to site large telescopes at the top of high mountains, where the air is thinner.

- Scientists can identify the chemical composition of stars by examining the light they emit with a spectrometer.

- Modern telescopes have sophisticated built-in spectrometers.

Analysing data

- When electromagnetic waves pass through the atmosphere, some wavelengths are absorbed much more than others.

- Radio waves are not absorbed at all by the atmosphere, so they can be used for communications.

- Gamma and X-rays are almost all absorbed by the atmosphere.

A graph showing the effect of the Earth's atmosphere on electromagnetic waves.

The Doppler effect

- As a siren on a police car passes, you will notice that the sound changes from high pitch as it approaches you to lower pitch as it moves away again.

- The sound waves are being squashed together as the car comes towards you and stretched as the car moves away. This is known as the **Doppler effect**.

- The Doppler effect can be observed for all waves, not just sound waves. For example, it can be observed in light emitted from a star.

- The wavelength of light decreases (and frequency increases) as a light source moves towards an observer. This is known as blue-shift.

- The wavelength of light increases (and frequency decreases) as a light source moves away from an observer. This is known as red-shift.

Remember!

The faster a source is moving, the greater the Doppler shift will be.

A moving source. What does it do to the waves?

Improve your grade

Positioning telescopes

Higher: Explain why it is best to site optical telescopes at the top of high mountains.   *AO1* [2 marks]

The life of stars

How stars are formed

- The Sun is our local star. It was formed about 4.6 billion years ago from a thin cloud of dust, gas (mostly hydrogen and helium) and ice, called a **nebula**.
- All the material in the nebula was pulled together by the **force of gravity**. It eventually started to spin and contract, and the temperature increased.
- At a few million °C, **nuclear fusion** reactions started to occur, the temperature rose even higher, and the new star began to emit its own light.

G–E

Fusion reactions

- The Sun is an average star, known as a **main sequence star**.
- Its surface temperature is about 5500 °C, which makes it appear yellow. The core temperature is about 14 million °C.
- The Sun's energy comes from nuclear fusion reactions. The vast amount of energy produced is emitted as electromagnetic radiation.
- The Sun is in a delicate balance between the force of gravity trying to pull all matter towards its centre and the force produced by the EM radiation pushing its way out.

D–C

- The nuclear fusion reaction in the Sun occurs when hydrogen nuclei fuse together to form a helium nucleus.
- The mass of the helium nucleus is slightly less than the total mass of the hydrogen nuclei. The lost mass becomes energy:
  hydrogen $\longrightarrow$ helium + energy

radiation force $\longrightarrow$
gravitational force $\longrightarrow$

Radiation force and gravitational force balance to keep the Sun from collapsing.

B–A*

Evolution of stars similar to the Sun

- Stars are born and eventually, after millions of years, they die. The journey of a star is known as its **life cycle**.
- Our Sun will eventually run out of hydrogen. When this happens, it will start to use helium as fuel to keep burning. This will produce different elements, such as carbon and oxygen.

nebula | Sun (main sequence) | red giant | white dwarf | black dwarf

The evolution of the Sun.

G–E

- The outer layers of the Sun (and similar-sized stars) will expand and cool, and it will become a **red giant**.
- Eventually, the core of the Sun will run out of elements to fuse and it will shrink under its own force of gravity. It will become very hot and dense – a **white dwarf**.
- The white dwarf will slowly cool. The star ends its life as a **black dwarf**.

D–C

Evolution of massive stars

- Larger, bluer stars were created from larger nebulae. Their evolution is different to that of stars like the Sun.
- The outer layers will expand and cool to form a **super red giant**.
- Nuclear fusion reactions in the star form heavier elements, such as iron and magnesium.
- When all the elements run out, the core will suddenly shrink and all the hot material will be ejected into space as a **supernova**. Very heavy elements will be formed.

large nebula | star | super red giant | supernova | black hole | neutron star

The evolution of a massive star.

B–A*

- The core shrinks to become a very dense **neutron star**. In extremely massive stars, a **black hole** is formed, which has such strong gravitational force that not even light can escape.

Improve your grade

Energy from the Sun
Higher: Explain where the Sun's energy comes from.

AO1 [4 marks]

Theories of the Universe

Steady State or Big Bang?

- The **Steady State theory** of the Universe was proposed by a group of scientists including Fred Hoyle in 1946. In this model the Universe:
  - is expanding
  - has unchanging density
  - spontaneously created matter, especially hydrogen, from empty space to maintain the same density
  - had no beginning and will never end.
- The **Big Bang** theory of the Universe was proposed by a group of scientists including George Gamov in 1948. In this model the Universe:
  - is expanding
  - is finite and ever-changing
  - was created about 14 billion years ago from an event called the Big Bang
  - may have an end, depending on its density.

> **Remember!**
> Both theories agree with the evidence that the Universe is expanding.

- Most scientists today believe the Big Bang theory is correct, because it explains the red-shift of light from distant galaxies and the existence of **cosmic microwave background** (CMB) **radiation**.

G–E

The expanding Universe

- CMB radiation is considered to be 'left-over radiation' from the Big Bang.
- It can be detected by radio telescopes and is the same strength in all directions.
- Dark lines can be detected against a continuous spectrum of colour in the Sun's spectrum. The lines are caused by the absorption of certain frequencies of light.
- The spectrum of light from distant stars show the same pattern, but the dark lines are shifted towards the red end of the spectrum – this is red-shift.
- The light from all distant galaxies is red-shifted. This means that galaxies are all moving away from us and from each other. The Universe is expanding.
- The further away the galaxy, the greater the red-shift. Therefore, the more distant galaxies are moving away more quickly.

D–C

Evidence for the Big Bang

- The fact that the Universe is expanding means that it must have started from the same point – the Big Bang.
- At the moment of this Big Bang, all matter expanded from a hot, dense singularity. After the first few seconds, hydrogen and helium were produced. As the Universe expanded, it cooled. Stars and galaxies formed.
- There is strong evidence for the Big Bang theory:
  - All galaxies show red-shift, which means the Universe must be expanding.
  - The temperature of the Universe was predicted to be –270 °C. This was confirmed by the COBE satellite in the 1990s.
  - The original radiation created in the Big Bang would have been gamma rays. Due to the expansion of the Universe, the Doppler effect has stretched the wavelength to become microwave radiation, with a wavelength of about 1 mm.
- The Steady State theory also accounts for the red-shift of galaxies, but it is rejected because it cannot explain the presence of CMB radiation or the abundance of light elements such as helium in the Universe.

B–A*

How science works

You should be able to:

- understand that nobody really knows how the Universe began – we only have theories and models
- explain how scientists are currently collecting more data to provide evidence for their theories
- describe how scientists may interpret the data in different ways.

Improve your grade

Steady State or Big Bang?

Foundation: The Steady State theory and the Big Bang theory were two opposing theories of the Universe in the 20th century.

Give **one** similarity between the two theories and **one** difference between them.   *AO1* [3 marks]

Ultrasound and infrasound

What you can and cannot hear

- Humans can hear sounds from 20 Hz up to about 20 000 Hz. Younger people can hear much higher-pitched sounds than older people.
- Sound with frequencies above 20 000 Hz is called **ultrasound**. Humans cannot hear ultrasound, but some animals can.
- Some animals, such as dolphins, use ultrasound for communication.
- Sound with frequencies lower than 20 Hz is called **infrasound**. Humans cannot hear infrasound, but they can feel the slow vibrations.
- Some animals, such as whales, use infrasound for communication.

Sound waves

- The speed of sound varies in different materials. The closer the particles in the material, the faster sound travels, so sound travels fastest in solids and slowest in gases.
- Ultrasound is used for foetal scanning, to safely 'see' inside the body.
- Different body tissues reflect ultrasound by differing amounts, and the echoes are used to create an image.
- Very high frequency ultrasound (about 1.5 MHz) is used to see fine detail.
- **Sonar** is a technique used by ships to determine the depth of water:
  - Pulses of ultrasound are reflected off the bottom of the seabed.
  - The time delay between sending the pulse and receiving the echo is used to calculate the distance travelled.

Ships use sonar to find the depth of water.

- Dolphins and bats use the same method to find their prey, but it is called **echolocation**.
- Infrasound is produced naturally by volcanoes, avalanches, ocean waves, hurricanes, earthquakes, meteorite explosions and animal movements.
- Infrasound is also produced by human activity, such as drilling for oil, and nuclear and chemical explosions.
- Infrasound can travel hundreds of kilometres through the Earth and atmosphere.
- Humans use infrasound to detect and monitor:
  - animal movements in remote locations
  - volcanic activity and meteorite strikes.

> **Remember!**
> Ultrasound is high-frequency sound. Infrasound is low-frequency sound.

Deep water

- With sonar, the speed of sound and the time delay can be used to calculate the depth of water:
  distance travelled by ultrasound pulse = speed of ultrasound × time delay
- Look at the diagram of the ship above. Remember – the ultrasound pulse has travelled from the ship to the seabed and back, so the depth is half the distance calculated.

Improve your grade

Measuring depth

Higher: A ship uses sonar to measure the depth of water as it approaches a harbour. A short pulse of ultrasound is sent out, and the time for it to return is measured as 20 milliseconds (a millisecond is a thousandth of a second).

If the speed of ultrasound is 1500 m/s, what is the depth of water? *AO2* [2 marks]

Earthquakes and seismic waves

The Earth and earthquakes

- The Earth is not a solid ball. It has several layers:
  - The inner core is very hot and dense solid iron.
  - The outer core is very hot, dense and liquid, mostly iron.
  - The mantle is hot, less dense and a mixture of solid and molten rock.
  - The crust is a very thin layer of solid rock.
- The crust floats on a hot liquid called **magma**. When magma comes to the surface from erupting volcanoes it is known as lava.

The layered Earth.

- The Earth's crust is split into sections known as **tectonic plates**, which move very slowly due to convection currents in the magma.
- Sudden movement of tectonic plates causes the shaking of the ground we call earthquakes:
  - As two plates try to slide past one another, friction between them at first prevents them from moving.
  - Eventually, the frictional forces cannot keep the plates still and they suddenly slip.
  - The immense energy stored in the compressed plates is released in the form of **seismic waves**.

- Alfred Wegener first suggested the movement of the Earth's crust in 1915. Since then, evidence has supported the idea:
  - Ancient rocks found in East Africa are identical to those found in South America.
  - Fossils found in both regions were from the same species of ancient aquatic reptile.

Seismic waves

- There are three types of seismic waves: P, S and L waves.

Wave	Type	Speed	Travel in solids	Travel in liquids
P waves	Longitudinal	Fast	Yes	Yes
S waves	Transverse	Slow	Yes	No
L waves	Combination	Very slow	Yes	Yes

- When an earthquake occurs on the seabed it can cause a tsunami (tidal wave). Earthquakes and tsunamis are unpredictable because:
  - Scientists cannot measure the pressure between the tectonic plates.
  - The fault lines are often deep within the Earth's crust.

A seismograph showing S, P and L waves.

- Scientists make predictions of the risk of earthquakes happening in a specific area, based on its earthquake history.
- Seismic waves are monitored around the world using **seismometers**. The trace from a seismometer is called a seismograph.
- The distance from the seismometer to the epicentre of the earthquake can be calculated from the time delay between receiving the P and S waves and their speeds.

EXAM TIP

You may be asked to investigate the unpredictability of earthquakes by measuring the force needed to move a block on a horizontal surface or the angle of tilt needed for a block to move on a slope. Learn these experiments and practise plotting the results on a graph.

- Transverse S waves cannot travel through the liquid outer core of the Earth, but they can travel in the solid crust and semi-solid mantle.
- Longitudinal P waves can travel through all the different layers of the Earth, but at differing speeds, so they **refract** at the boundaries.
- Both P and S waves travel faster in more dense materials, so the deeper inside the Earth –where there is very high pressure – the more dense the material and the faster both waves travel. This causes the waves to refract and follow curved paths.
- Both P and S waves will **reflect** at the boundaries between the layers.

S and P waves through the Earth. Note that there are no S waves at the core.

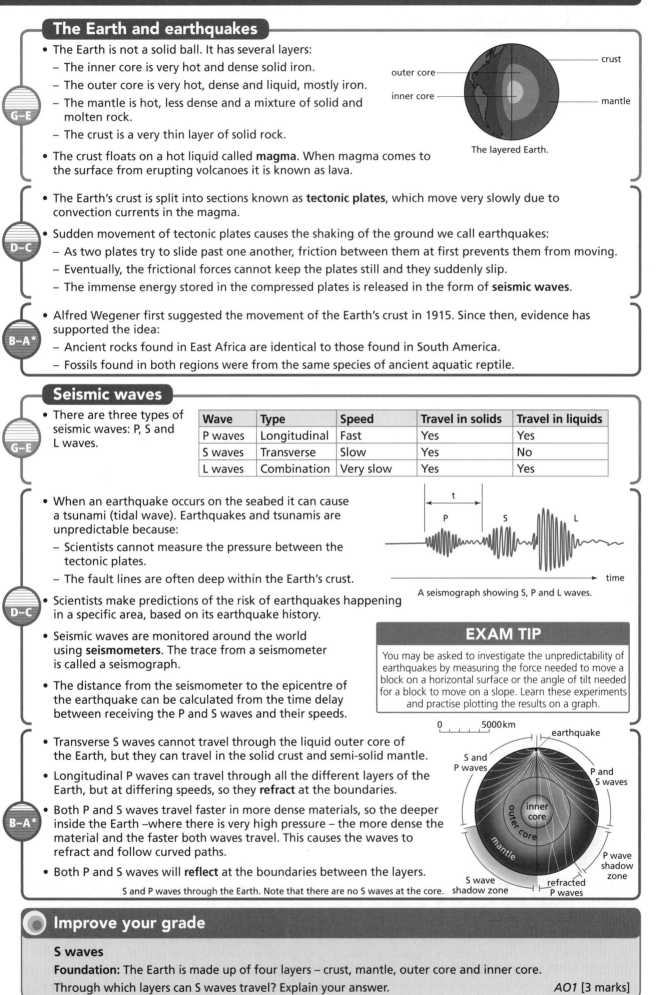

Improve your grade

S waves

Foundation: The Earth is made up of four layers – crust, mantle, outer core and inner core.
Through which layers can S waves travel? Explain your answer.   *AO1* [3 marks]

Electrical circuits

Charge and current

- Circuit diagrams are used to show electrical circuits. When there is a complete closed circuit, an electric **current** will flow.

- Electric current is a *flow* of **charge**. In copper wires, the charge is due to negatively charged **electrons**.

- Electric current is measured in **amperes** (A), using an **ammeter** connected in **series**.

- The electrons in the circuit are attracted to the positive terminal of the battery, but **conventional current** flows from positive to negative.

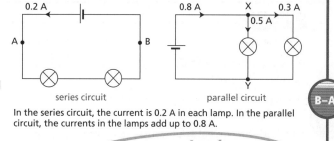

A circuit diagram showing an ammeter connected in series.

Types of circuits

- In series circuits, components are connected end to end in a single loop. The current is the same all the way round the series circuit. An ammeter placed at points A and B will show the same reading.

- In **parallel** circuits, the components are connected across each other. The current will split at the junction X and re-join at junction Y.

- The unit of electric charge is the **coulomb** (C). An electron carries a tiny amount of charge – only -1.6×10^{-19} C.

$$\text{electric current (A)} = \frac{\text{electric charge (C)}}{\text{time (s)}}$$

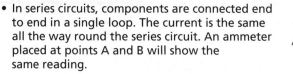

series circuit   parallel circuit

In the series circuit, the current is 0.2 A in each lamp. In the parallel circuit, the currents in the lamps add up to 0.8 A.

Remember!
Electric current is the rate of flow of charge. A current of 1 A is 1 C of charge per second.

Voltage

- The cell or battery in a circuit gives energy to the electrons so that they can transfer energy to the components in the circuit.

- The chemical energy in the cell is converted to electrical energy of the electrons.

- A **voltmeter** measures the amount of energy transferred in a component. It is connected in parallel across the component.

- The **voltage**, or **potential difference**, is measure in **volts**.

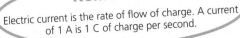

How is the voltmeter connected to the lamp?

- In series circuits, the voltage across individual components adds up to the voltage across the power supply: $V = V_1 + V_2$

- In parallel circuits, the voltage across each component is the same as the voltage across the power supply: $V = V_1 = V_2$

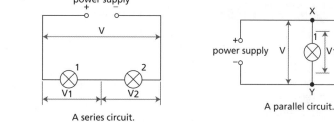

A series circuit.   A parallel circuit.

- Potential difference is the amount of energy per unit charge. A p.d. of 1 volt is 1 joule of energy per coulomb of charge.

$$\text{potential difference (volt)} = \frac{\text{energy (joule)}}{\text{charge (coulomb)}}$$

Improve your grade

Potential difference

Higher: A cell has a potential difference of 1.5 V.

What is meant by the term 'potential difference'?   *AO1* [3 marks]

Electrical power

Calculating power

- **Electrical power** is the rate of energy transfer. It is measured in **watts** (W). One watt is one joule per second.

  power (W) = $\dfrac{\text{energy transferred (J)}}{\text{time (s)}}$ or $P = \dfrac{E}{t}$

- All electrical appliances are marked with a power rating in watts or kilowatts.

- In the circuit opposite, the voltmeter reads 12 V and the ammeter reads 2.0 A. The power of the lamp can be worked out as: 12 V × 2.0 A = 24 W

  electrical power (W) = current (A) × potential difference (V) or $P = I \times V$

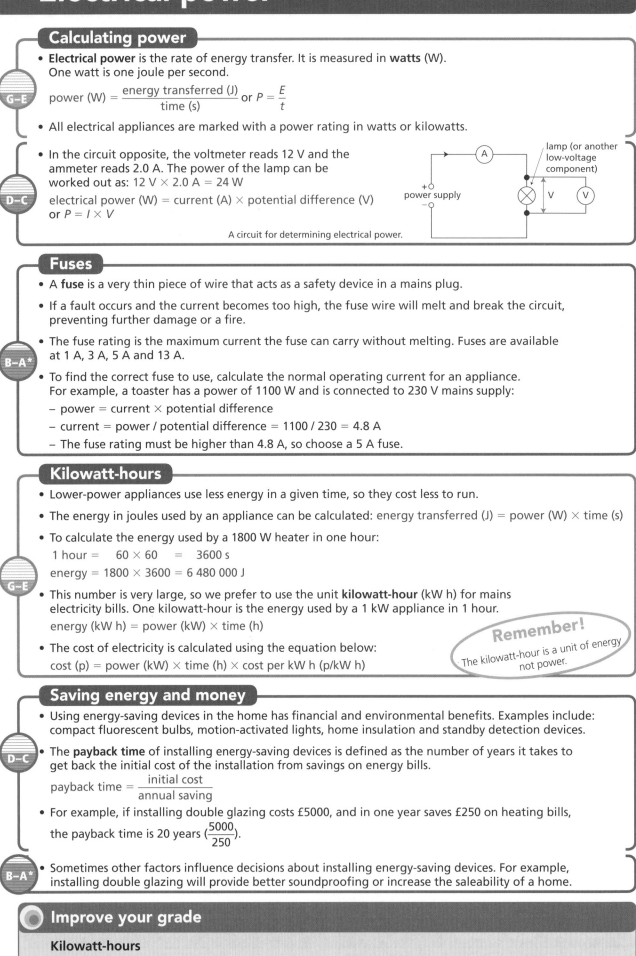

A circuit for determining electrical power.

Fuses

- A **fuse** is a very thin piece of wire that acts as a safety device in a mains plug.

- If a fault occurs and the current becomes too high, the fuse wire will melt and break the circuit, preventing further damage or a fire.

- The fuse rating is the maximum current the fuse can carry without melting. Fuses are available at 1 A, 3 A, 5 A and 13 A.

- To find the correct fuse to use, calculate the normal operating current for an appliance. For example, a toaster has a power of 1100 W and is connected to 230 V mains supply:
  - power = current × potential difference
  - current = power / potential difference = 1100 / 230 = 4.8 A
  - The fuse rating must be higher than 4.8 A, so choose a 5 A fuse.

Kilowatt-hours

- Lower-power appliances use less energy in a given time, so they cost less to run.

- The energy in joules used by an appliance can be calculated: energy transferred (J) = power (W) × time (s)

- To calculate the energy used by a 1800 W heater in one hour:

  1 hour =   60 × 60   =   3600 s

  energy = 1800 × 3600 = 6 480 000 J

- This number is very large, so we prefer to use the unit **kilowatt-hour** (kW h) for mains electricity bills. One kilowatt-hour is the energy used by a 1 kW appliance in 1 hour.

  energy (kW h) = power (kW) × time (h)

- The cost of electricity is calculated using the equation below:

  cost (p) = power (kW) × time (h) × cost per kW h (p/kW h)

Remember! The kilowatt-hour is a unit of energy not power.

Saving energy and money

- Using energy-saving devices in the home has financial and environmental benefits. Examples include: compact fluorescent bulbs, motion-activated lights, home insulation and standby detection devices.

- The **payback time** of installing energy-saving devices is defined as the number of years it takes to get back the initial cost of the installation from savings on energy bills.

  payback time = $\dfrac{\text{initial cost}}{\text{annual saving}}$

- For example, if installing double glazing costs £5000, and in one year saves £250 on heating bills, the payback time is 20 years ($\dfrac{5000}{250}$).

- Sometimes other factors influence decisions about installing energy-saving devices. For example, installing double glazing will provide better soundproofing or increase the saleability of a home.

Improve your grade

Kilowatt-hours

Foundation: An oven has a power rating of 2000 W and it takes 45 minutes to cook a cake. The cost of 1 kW h of electrical energy is 22 p.

What is the cost of the electricity to cook the cake? *AO2 [2 marks]*

Energy resources

Non-renewable energy resources

- Coal, oil and natural gas are called fossil fuels because they formed from the remains of prehistoric plants and animals.
- Fossil fuels and nuclear fuel are **non-renewable resources** because they cannot be replaced and will eventually run out.

G–E

Resource	Advantages	Disadvantages
Coal	Fuel is cheap; coal-burning power stations have a quick start-up time; coal will last at least 200 years.	Burning coal releases CO_2 and SO_2; SO_2 produces acid rain; mining can be dangerous; stockpiles are needed to meet demand.
Natural gas	Gas-fired power stations are efficient and have the quickest start-up time, so they are flexible at meeting demand for power; gas will last another 50 years; gas does not produce SO_2.	Burning gas releases CO_2 (although less than coal and oil); pipelines necessary for transporting gas are expensive.
Oil	Oil-burning power stations have a quick start-up time; there is enough oil left to last 50 years.	Burning oil releases CO_2 and SO_2; oil prices are extremely variable; there is a danger of spillage and pollution during transport of oil by road, rail or sea.
Nuclear	Nuclear power stations are located away from population centres; it does not produce CO_2 or SO_2.	Building and decommissioning a power station is expensive; start-up time is the longest; radioactive waste remains dangerous for thousands of years.

D–C

The greenhouse effect

- Infrared radiation emitted by the surface of the Earth is absorbed by **greenhouse gases** in the atmosphere. This causes the atmosphere to warm up in a process called the **greenhouse effect**.
- Burning fossil fuels increases the levels of carbon dioxide in the atmosphere. Some scientists believe that this will cause the average temperature of the Earth to rise in the future.

B–A*

Renewable energy resources

- **Renewable energy resources** are also used to generate electricity:
  - Solar power turns light energy from the Sun into electrical energy, using solar cells.
  - Wind power is used to rotate huge propeller blades on a turbine to generate electricity.
  - Wave power is generated when large floats containing coils and magnets move up and down with ocean waves.
  - Hydroelectric power is when fast-flowing water, stored in a reservoir above a power station, is used to generate electricity.
  - Tidal power uses seawater from incoming and outgoing tides to create electricity.
  - Biomass is organic material from decaying plant or animal waste that can be used as a fuel in a power station, in the same way as fossil fuels are. Wood can be used in a power station in a similar way.

G–E

- Most renewable sources cost nothing to use, and produce no greenhouse gases, but the cost of building renewable power stations is substantial. Other advantages and disadvantages of renewable resources are listed in the table below.

Resource	Advantages	Disadvantages
Solar	Useful in remote areas; single homes can have their own electricity supply.	No power at night or when cloudy.
Wind	Can be built offshore.	Can cause noise and visual pollution; amount of electricity depends on the weather.
Wave and tidal	Ideal for island countries.	May be opposed by local or environmental groups.
Hydroelectric	Creates water reserves as well as electricity supplies.	Can cause flooding of surrounding communities and landscapes.
Biomass and wood	Cheap and readily available source of energy; if replaced, biomass can be a long-term, sustainable energy source.	Gives off CO_2 and SO_2 when burnt; biomass and wood are only renewable resources if crops and trees are replanted.

D–C

Improve your grade

Understanding resources

Foundation: Fuels are burnt in power stations. Fossil fuels are non-renewable energy sources, and biomass is a renewable energy source.

Explain what is meant by the terms 'non-renewable' and 'renewable'.　　　　　*AO1* [2 marks]

Generating and transmitting electricity

Inducing a current

G–E

- A voltage is **induced** when a wire is moved in a magnetic field. If the piece of wire is part of a circuit, a current will flow. This is called electromagnetic induction.
- The direction of the current is reversed when the motion of the wire is reversed, or when the magnet is turned round.
- A voltage is always induced when there is relative movement between a magnet and a coil of wire. The induced voltage is larger when the magnet is moved more quickly.

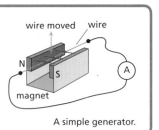

A simple generator.

Generators

D–C

- In an electrical **generator**, a coil is rotated in a magnetic field. As the coil rotates, it cuts the magnetic field lines to induce a voltage across the coil.
- The coil has slip rings, which are connected to a circuit via brushes. This causes an alternating current to flow in the circuit.
- The current will increase if the speed of motion increases, if a stronger magnet is used, or if there are more turns of wire in the coil.
- **Direct current** (d.c.) always flows in the same direction. A cell provides direct current.
- **Alternating current** (a.c.) changes direction at a frequency determined by the rotating coil.

B–A*

- Mains electricity is generated at a frequency of 50 Hz.
- The variation of current with time follows a sinusoidal curve.

A simple generator.

Typical alternating current from a mains generator.

Transformers

G–E

- A **transformer** is a device that changes the size of an alternating voltage.
- Two coils of wire are wrapped round a soft iron core. The alternating voltage supply is connected to the **primary coil**, and the output alternating voltage is induced across the **secondary coil**.
- A **step-up transformer** converts a low voltage input to a higher voltage output. The primary coil will have fewer turns than the secondary coil.
- A **step-down transformer** converts a high voltage input to a lower voltage output. The primary coil will have more turns than the secondary coil.

A simple transformer.

D–C

- The **National Grid** is the network of pylons and cables that transmit electrical energy from power stations to users across the UK.
- Transformers are used in the National Grid:
  - Step-up transformers increase the 23 kV power-station output to higher voltages (275 kV or 400 kV). This reduces heat loss in the cables and improves efficiency of transmission.
  - Step-down transformers lower the voltage to 230 V, which is safer for homes.
- Electricity transmission is hazardous. Overhead cables are not insulated, so can cause serious harm or even death if touched.
- Risks can be minimised by burying cables underground, but this is expensive and makes it difficult to maintain the cables. Suspending cables from pylons is a cheaper option.

Remember!
Transformers only work with a.c. electricity.

B–A*

- The a.c. voltage in the primary coil of a transformer creates an ever-changing magnetic field around it.
- The magnetic soft-iron core channels the magnetic field through the secondary coil.
- The alternating magnetic field will continuously cut through the wires in the secondary coil and an a.c. voltage will be induced.
- The turns ratio is equal to the voltage ratio:

$$\frac{\text{output voltage at secondary coil}}{\text{input voltage at primary coil}} = \frac{\text{number of turns on secondary coil}}{\text{number of turns on primary coil}}$$

Improve your grade

Transformers

Foundation: Describe how step-up transformers are used in the National Grid. *AO1* [4 marks]

Energy and efficiency

Types of energy

• Energy exists in many forms, as summarised in the table below.

Name	Description	Examples
Thermal (heat) energy	An object at a higher temperature has greater thermal energy	Heater; the Sun; hot water
Light energy	Light is a wave that is emitted from anything at a very high temperature	Lamp; stars; fire
Electrical energy	This is usually associated with electric current	Mains supply; overhead cables; output from a transformer
Sound energy	An object vibrating will emit sound	Buzzer; bell; siren; person talking
Kinetic energy	This is energy due to movement	Person running; high-speed train; planet orbiting the Sun
Chemical energy	This is energy stored by atoms	Food; chemical cell; coal
Nuclear energy	This is energy stored by the nuclei of atoms	Nuclear power station; radioactivity; nuclear bombs
Elastic potential energy	An object that is pulled or squashed has this type of energy	A stretched rubber band; cables supporting a bridge
Gravitational potential energy	This is energy due to an object's position in the Earth's gravitational field. An object lifted higher will have greater gravitational potential energy	Aeroplane in the sky; person up a ladder

G–E

Energy transfers

• Whenever anything moves or changes, an energy transfer must happen.

• For example, in a firework, the chemical energy stored in the firework becomes heat, light and sound when the firework explodes.

chemical energy ⟶ thermal energy + light energy + sound energy

• Energy cannot be created or destroyed. It can only be transferred from one form to another. This is known as the **principle of conservation of energy**.

D–C

Remember!
The total energy input must equal the total energy output.

Efficiency of devices

• In all energy transfers, some of the energy output is useful and some is unwanted or wasted energy. For example, in a light bulb the useful energy output is the light, and the wasted energy output is the heat.

• In most devices heat energy is wasted. The heat energy produced may be due to electric currents flowing in wires or due to friction between moving surfaces.

G–E

• **Efficiency** is a measure of how well a device transfers energy in the form we want. Efficiency is calculated using the equation:

$$\text{efficiency} = \frac{\text{(useful energy transferred by the device)}}{\text{(total energy supplied by the device)}} \times 100\%$$

• **Sankey diagrams** are energy-transfer diagrams that show the different forms energy takes during a transfer. The thickness of the arrow in the diagram is drawn to scale to show the amount of energy transferred.

D–C

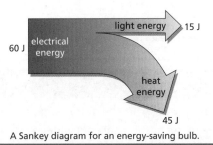

A Sankey diagram for an energy-saving bulb.

Improve your grade

Calculating efficiency

Higher: An electric drill transfers every 50 J of electricity into 20 J of useful kinetic energy.

Calculate the percentage efficiency of the drill.

What forms does the wasted energy in the transfer take?

AO2 [4 marks]

Radiated and absorbed energy

Infrared radiation

- All objects can emit infrared radiation, absorb infrared radiation and reflect infrared radiation.
- The amount of infrared radiation emitted by an object depends on its temperature and its surface.
- All objects at temperatures greater than absolute zero emit (or radiate) infrared radiation. The hotter an object, the more power it radiates.
- A dull black surface loses energy more quickly than a bright shiny surface because:
  - a dull black surface is a good radiator of heat
  - a bright shiny surface is a poor radiator of heat.
- A dull black surface is also a good absorber of heat radiation.
- A bright shiny surface is a poor absorber, as it reflects the heat radiation away.
- Objects that are warmer than their surroundings will emit more energy per unit time than they absorb.
- Objects that are cooler than their surroundings will absorb more energy per unit time than they emit.

Remember!
Dull black surfaces are good emitters *and* good absorbers.

Staying at the same temperature

- All objects are continually absorbing and emitting heat radiation.
- When the rate of absorption is greater than the rate of radiation, the temperature will rise (and vice versa).
- Eventually, the rate of heat absorption will equal the rate of heat radiation and the temperature remains steady. This is called thermal equilibrium.
- For example, the hot filament in a bulb will reach a steady temperature when the input of electrical power is equal to the power radiated away from the filament.

Experimenting with radiation

- All atoms have both **kinetic energy** (due to their vibrations) and **potential energy** (due to their position).
- When substances are heated, the temperature increases and the atoms gain more kinetic energy.
- The heat energy of a substance is the total kinetic energy and potential energy of all the atoms in the substance.
- To investigate the different rates of heat radiation of different surfaces, you can compare the temperature over time of a silvered beaker and a blackened beaker (see below).
- To investigate the different rates of heat absorption, you can use a radiant heater between two plates – one silvered and one blackened. Then time how long it takes for each plate to get hot enough to melt some wax.

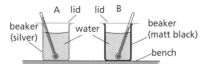

A cooling experiment with shiny and black beakers. The thermometers in each beaker will indicate the amount of thermal energy radiated.

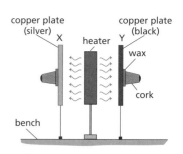

An absorption experiment with shiny and black metal plates. The melting wax is an indicator of the amount of thermal energy absorbed.

How science works

In the investigations mentioned above, you should be able to:
- ensure that there is a valid test – there must be only one independent variable
- repeat readings to increase the reliability of the investigation.

Improve your grade

Staying cool
Foundation: In many Mediterranean countries, such as Greece, the houses are all painted white. Explain why this is, using ideas about heat radiation. *AO2* [3 marks]

P1 Summary

Light is refracted and reflected at surfaces and boundaries between different materials.

Refracting telescopes use an objective lens to form a real image, which is then magnified. Reflecting telescopes use a concave mirror, which gives a clearer image and allows larger telescopes to be made.

Visible light and the solar system

The model of the Universe changed from a geocentric to a heliocentric model.

Observations of the Universe in the past used light, but nowadays many different types of wave are used to gain more evidence.

Waves (transverse or longitudinal) transmit energy but not matter. They are described in terms of wavelength, frequency and amplitude.

The EM spectrum is a family of waves that all travel at the same speed in a vacuum. They are: radio waves, microwaves, infrared, visible light, ultraviolet, X-rays and gamma rays.

The electromagnetic spectrum

Ionising alpha, beta and gamma radiation comes from the nucleus of radioactive elements. The higher the frequency of the EM waves, the higher the ionising power.

The Solar System is our group of planets, moons, asteroids and comets that orbit the Sun.

Scientists study the Universe with all parts of the EM spectrum.

Waves and the Universe

Nuclear reactions take place within all stars, which give out vast amounts of energy in the form of EM waves.

The human hearing range is approximately 20 Hz to 20 000 Hz.
Frequencies lower than 20 Hz are called infrasound.
Frequencies higher than 20 000 Hz are called ultrasound.

Waves and the Earth

Seismic waves are generated by earthquakes. S waves and P waves have different properties.

The Earth's crust is made up of tectonic plates, which are constantly moving due to convection currents in the mantle. Earthquakes are caused by the relative movement of the plates at plate boundaries.

Power is the rate at which energy is used. Power (W) = energy (J)/time (s).
Electrical power (W) = current (A) x potential difference (V)

Fossil fuels are non-renewable resources. Burning them and contributes to the greenhouse effect.

Renewable energy resources include: wind, waves, tidal, hydroelectric and solar.

Generation and transmission of electricity

Electromagnetic induction occurs when there is relative motion between a wire and a magnetic field. An electric current is generated.

Electricity generated in power stations is alternating current (a.c.). Electricity from a battery is direct current (d.c.).

Transformers are used to change the voltage of a.c. electricity.

Energy cannot be created or destroyed – only converted from one form to another, as depicted in Sankey diagrams.
In all energy transfers, some energy is wasted as heat.

Energy and the future

All objects at temperatures above absolute zero emit infrared radiation.

Electrostatics

The structure of atoms

- All matter is made up of tiny **atoms**. Each atom is about 0.0000002 mm in diameter.
- Scientists believe in the nuclear model of the atom:
  - The **nucleus** is about 100 000 times smaller than the diameter of the atom.
  - The nucleus contains particles called **protons** and **neutrons**.
  - Protons are positively charged and neutrons have no charge, so the nucleus is positive.
  - Negative **electrons** occupy the space around the nucleus.
- The table below shows the properties of atomic particles. The mass and charge are shown relative to that of the proton.

Particle	Where found	Relative mass	Relative charge
Neutron	Inside the nucleus	1	0
Proton	Inside the nucleus	1	+1
Electron	Outside the nucleus	0	−1

- If an atom loses electrons it will have an overall positive charge; if it gains electrons it will have an overall negative charge.
- A charged atom is called an **ion**.

Static electricity

- An **insulator** can be charged by **friction**. For example, if you rub a balloon against your clothes and then hold it against a wall, the balloon sticks to the wall.
  - The friction transfers electrons to the balloon, making it negatively charged.
  - The charged balloon repels some of the electrons away from the surface of the wall, leaving the surface of the wall with a positive charge.
  - Opposite charges attract, so the balloon is attracted to the wall.
- The separated charges in the wall are called induced charges.
- Some other examples of electrostatic phenomena are:
  - A stream of water can be bent towards a charged insulator.
  - Synthetic clothing clings to your body.
  - A comb becomes charged when you comb your hair or rub it with a cloth. The comb will attract your hair or small pieces of paper.
- Simply rubbing a glass rod with a woollen cloth will cause a transfer of charge. The wool gains electrons to become negatively charged. The glass loses electrons and is left with an equal positive charge. This is called static charge.
- All insulators can be charged by friction:
  - Some insulators lose electrons by friction. They become positively charged.
  - Other insulators gain electrons by friction and become negatively charged.
  - Each insulator acquires an equal but opposite charge.

> **Remember!**
> Charged objects exert a force on each other. The force depends on the type of charge each one has: like charges repel and unlike charges attract.

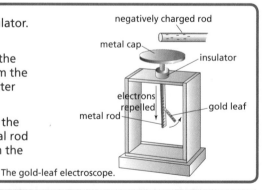

What is the net charge on the rod and cloth?

Gold-leaf electroscope

- A gold-leaf electroscope is used to find the charge on an insulator. A metal rod connects a metal cap to gold leaf.
- Bringing a negatively charged rod close to the cap will repel the electrons down the metal rod. The gold leaf moves away from the metal rod because they both have the same charge. The greater the charge, the further away the leaf moves.
- Bringing a positively charged rod close to the cap will attract the electrons towards the cap, leaving the gold leaf and the metal rod with a positive charge. Again, the gold leaf moves away from the metal rod because they both have the same charge.

The gold-leaf electroscope.

Improve your grade

Forces and attraction

Foundation: Daisy combs her hair with a plastic comb. She can then use the plastic comb to pick up small pieces of paper.

Explain why the paper is attracted to the comb.

AO1 [3 marks]

Uses and dangers of electrostatics

Uses of electrostatics

- Electrostatic paint sprayers are used on many metal objects:
  - The object to be sprayed, such as a car, is connected to a negative supply.
  - The sprayer charges the tiny droplets of paint as they emerge from the nozzle.
  - The charged droplets of paint repel each other and spread out to form a dispersed cloud.
  - The positive droplets are attracted to the negatively charged object being sprayed.

- The advantages to using this technique are that:
  - less paint is used
  - the object is given an even coating of paint
  - every part of the object attracts the paint, even the underside.

- The same technique is used to spray plants with **insecticides**. The insecticide is given a positive charge and the plants acquire a negative charge by induction.

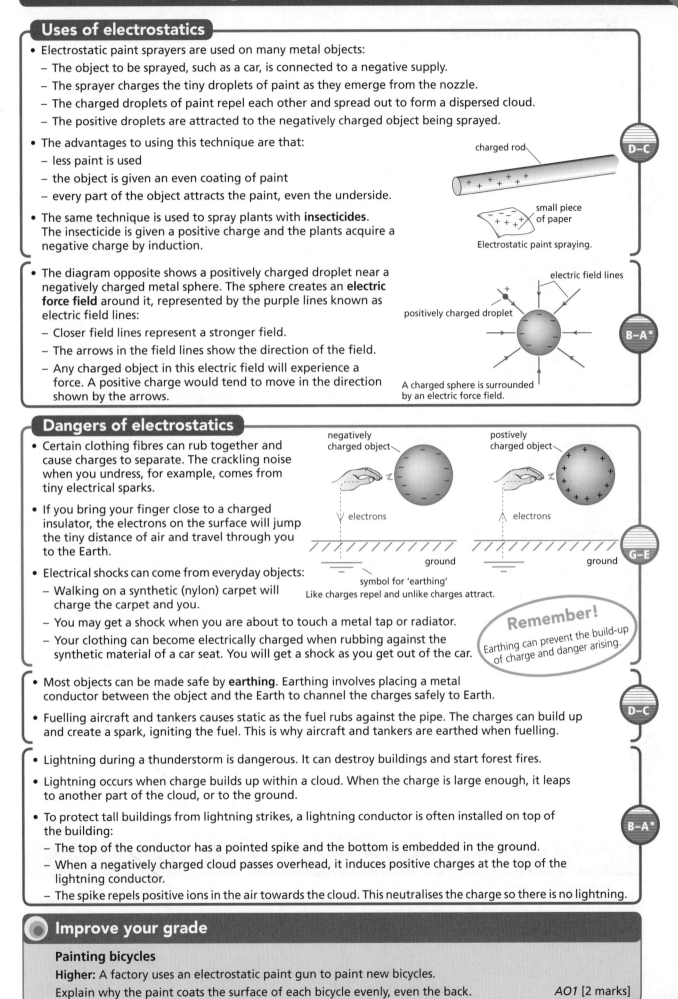

charged rod

small piece of paper

Electrostatic paint spraying.

D–C

- The diagram opposite shows a positively charged droplet near a negatively charged metal sphere. The sphere creates an **electric force field** around it, represented by the purple lines known as electric field lines:
  - Closer field lines represent a stronger field.
  - The arrows in the field lines show the direction of the field.
  - Any charged object in this electric field will experience a force. A positive charge would tend to move in the direction shown by the arrows.

electric field lines

positively charged droplet

A charged sphere is surrounded by an electric force field.

B–A*

Dangers of electrostatics

- Certain clothing fibres can rub together and cause charges to separate. The crackling noise when you undress, for example, comes from tiny electrical sparks.

- If you bring your finger close to a charged insulator, the electrons on the surface will jump the tiny distance of air and travel through you to the Earth.

- Electrical shocks can come from everyday objects:
  - Walking on a synthetic (nylon) carpet will charge the carpet and you.
  - You may get a shock when you are about to touch a metal tap or radiator.
  - Your clothing can become electrically charged when rubbing against the synthetic material of a car seat. You will get a shock as you get out of the car.

negatively charged object

postively charged object

electrons

electrons

ground

ground

symbol for 'earthing'

Like charges repel and unlike charges attract.

G–E

Remember!
Earthing can prevent the build-up of charge and danger arising.

- Most objects can be made safe by **earthing**. Earthing involves placing a metal conductor between the object and the Earth to channel the charges safely to Earth.

- Fuelling aircraft and tankers causes static as the fuel rubs against the pipe. The charges can build up and create a spark, igniting the fuel. This is why aircraft and tankers are earthed when fuelling.

D–C

- Lightning during a thunderstorm is dangerous. It can destroy buildings and start forest fires.

- Lightning occurs when charge builds up within a cloud. When the charge is large enough, it leaps to another part of the cloud, or to the ground.

- To protect tall buildings from lightning strikes, a lightning conductor is often installed on top of the building:
  - The top of the conductor has a pointed spike and the bottom is embedded in the ground.
  - When a negatively charged cloud passes overhead, it induces positive charges at the top of the lightning conductor.
  - The spike repels positive ions in the air towards the cloud. This neutralises the charge so there is no lightning.

B–A*

Improve your grade

Painting bicycles

Higher: A factory uses an electrostatic paint gun to paint new bicycles.

Explain why the paint coats the surface of each bicycle evenly, even the back.   *AO1 [2 marks]*

Current, voltage and resistance

Charge and current

G–E

- **Electric current** is the rate of flow of charge. It is measured with an **ammeter** (connected in **series**) and its unit is **amperes** or 'amps' (A).
- A cell is a chemical device with its own positive and negative terminals which push electrons around a circuit. A battery is a collection of cells, often joined together in series.
- A cell or a battery provides a **direct current** (d.c.). This means that the electrons travel in one direction only.

D–C

- Charge is measured in **coulombs** (C).
- In metals, current is due to the flow of electrons. Each electron has a tiny negative charge of -1.6×10^{-19} C.
- The current is 1 ampere when the rate of flow of charge is 1 coulomb per second.
- Current and charge are linked by the following equation, where Q is the charge in coulombs, I is the current in amperes and t is the time in seconds: $Q = I \times t$

Resistance

G–E

- Components have difference resistances, and can be used to control the size and current in circuits.
- The diagram opposite shows components (lamps) connected in series.
  - The current in a series circuit is the same all the way round.
  - The current will be larger when more cells are used, but the ammeters will still show the same current at all points around the circuit.
- The diagram opposite shows components connected in **parallel**.
  - The current splits at the junction J.
  - Electric current is conserved at the junction.
- Current is conserved at a junction because the total number of electrons entering a junction must be equal to the total number of electrons leaving the junction.

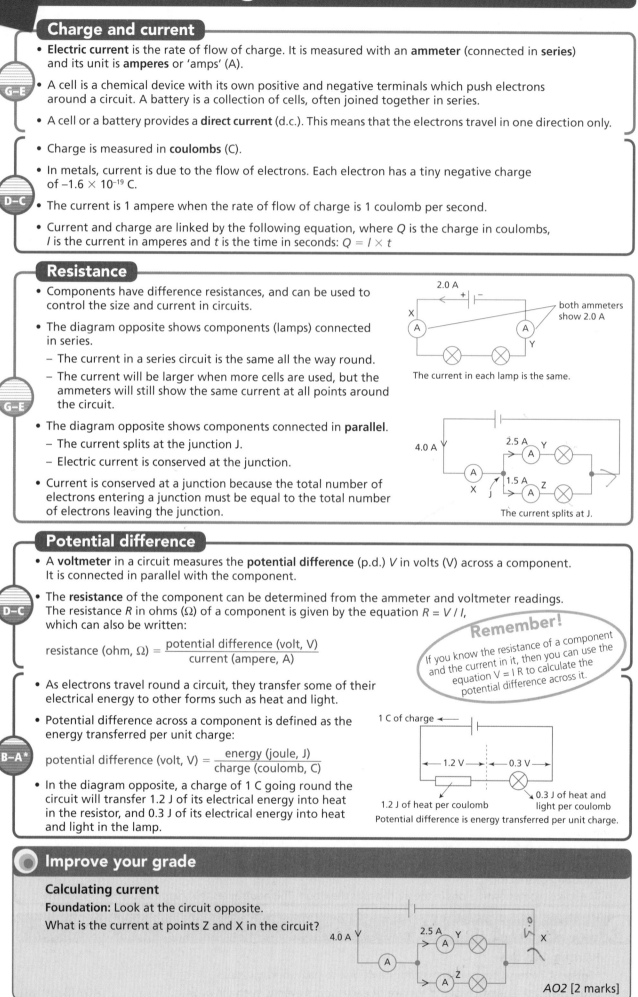

The current in each lamp is the same.

The current splits at J.

Potential difference

D–C

- A **voltmeter** in a circuit measures the **potential difference** (p.d.) V in volts (V) across a component. It is connected in parallel with the component.
- The **resistance** of the component can be determined from the ammeter and voltmeter readings. The resistance R in ohms (Ω) of a component is given by the equation $R = V / I$, which can also be written:

$$\text{resistance (ohm, } \Omega) = \frac{\text{potential difference (volt, V)}}{\text{current (ampere, A)}}$$

Remember!
If you know the resistance of a component and the current in it, then you can use the equation V = I R to calculate the potential difference across it.

- As electrons travel round a circuit, they transfer some of their electrical energy to other forms such as heat and light.

B–A*

- Potential difference across a component is defined as the energy transferred per unit charge:

$$\text{potential difference (volt, V)} = \frac{\text{energy (joule, J)}}{\text{charge (coulomb, C)}}$$

- In the diagram opposite, a charge of 1 C going round the circuit will transfer 1.2 J of its electrical energy into heat in the resistor, and 0.3 J of its electrical energy into heat and light in the lamp.

1 C of charge

1.2 V → ← 0.3 V

1.2 J of heat per coulomb

0.3 J of heat and light per coulomb

Potential difference is energy transferred per unit charge.

Improve your grade

Calculating current
Foundation: Look at the circuit opposite.
What is the current at points Z and X in the circuit?

AO2 [2 marks]

Unit P2

Topics 1 and 2: 1.9, 1.10, 1.11, 1.12, 1.13, 2.1, 2.2, 2.3, 2.4, 2.5, 2.8

Lamps, resistors and diodes

Changing currents in circuits

- The potential difference *V* across the wire and the current *I* in the wire can be used to determine its resistance.

- The resistance of the wire is directly proportional to its length. For example, the resistance doubles when the length is doubled.

- The resistance of a variable resistor (or **rheostat**) can be changed manually.

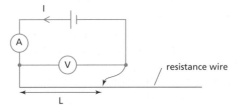

The meters help you to determine the resistance of the wire.

- The diagram below shows a rheostat used in a circuit to change the brightness of a lamp. When the resistance of the variable resistor is set:
  - to its maximum value, the current in the circuit is low and the lamp is dimly lit
  - to its lowest value (often zero), the current in the circuit is high and the lamp is fully lit.

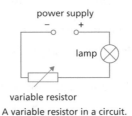

A variable resistor in a circuit.

G–E

Current against potential difference graphs

- Circuits can be used to investigate how a current *I* varies with potential difference *V* for devices such as a **filament lamp**, a **diode** and a **fixed resistor**. In the diagram opposite, the current is altered using the variable resistor.

- The graphs below show the *I* against *V* for a filament lamp, a 100 W fixed resistor and a diode.
  - In the filament lamp the current increases with p.d. As the current increases, the filament gets hotter and the resistance increases.
  - In the fixed resistor, the current is directly proportional to the p.d. Its resistance is constant.
  - The diode only conducts in one direction. It has an infinite resistance when the current is zero. The diode has low resistance when it conducts.

A circuit for investigating any component.

D–C

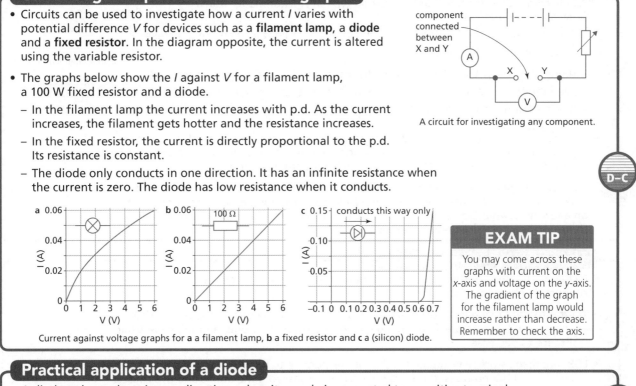

Current against voltage graphs for **a** a filament lamp, **b** a fixed resistor and **c** a (silicon) diode.

EXAM TIP

You may come across these graphs with current on the *x*-axis and voltage on the *y*-axis. The gradient of the graph for the filament lamp would increase rather than decrease. Remember to check the axis.

Practical application of a diode

- A diode only conducts in one direction, when its anode is connected to a positive terminal and its cathode is connected to a negative terminal.

- A diode and a lamp can be used to work out the polarities of an unmarked cell or power supply.

B–A*

Improve your grade

Variable resistors

Foundation: Explain how a circuit with a variable resistor can be used to control the brightness of a lamp in the same circuit.

AO2 [3 marks]

Heating effects, LDRs and thermistors

Heating effect of electric current

G–E
- The current in a resistor causes it to heat up. This heat is made use of in electric heaters and the heating element of an electric kettle.
- Excessive heat produced in electrical devices can cause fires if there are no safety features. This is why you need a cooling fan in a laptop. An overloaded mains socket is a potential fire hazard.

D–C
- The **electrical power** of a device can be used to calculate how much electrical energy it will transfer and how much it would cost to use the device.
- The power of a device is related to the current it carries and the potential difference (p.d.) across it. Power can be calculated using the equation $P = I \times V$, where P is the power, I is the current and V is the p.d. You can also write this equation as:

$$\underset{\text{(watt, W)}}{\text{electrical power}} = \underset{\text{(ampere, A)}}{\text{current}} \times \underset{\text{(volt, V)}}{\text{potential difference}}$$

- The energy transferred by any component can be found by multiplying the power by the time:

$$\underset{\text{(joule, J)}}{\text{energy transferred}} = \underset{\text{(watt, W)}}{\text{power}} \times \underset{\text{(second, s)}}{\text{time}}$$

or:

$$\underset{\text{(joule, J)}}{\text{energy transferred}} = \underset{\text{(ampere, A)}}{\text{current}} \times \underset{\text{(volt, V)}}{\text{potential difference}} \times \underset{\text{(second, s)}}{\text{time}}$$

Remember!
You can find the current and the voltage from the ammeter and voltmeter readings of a device in a circuit, and use the values to calculate the power and the energy transfer.

B–A*
- When current flows, electrons travel through the **lattice** of a metal.
- Electric current causes heating. The heating is the result of collisions between electrons and the ions in the lattice.
- The collisions cause increased vibrations (around fixed positions) of the ions. This is what we mean by heat energy.

LDRs and thermistors

G–E
- A **light-dependent resistor** (LDR) is a special type of resistor made from a **semiconductor** material.
  - The resistance of an LDR decreases as the intensity of light incident on it increases.
  - More light means less resistance.
- A **thermistor** is a special type of resistor. Its resistance depends on its temperature.
  - The resistance of a thermistor decreases as its temperature increases.
  - Greater temperature means less resistance.

Light-dependent resistor

Thermistor

D–C
- Circuits with LDRs can be used to detect changes in light **intensity**. Examples of where LDRs are used include simple light-meters, light-sensitive circuits in burglar alarms, and circuits for automatically controlling how much light enters a camera.
- Circuits with thermistors can be used to detect changes in temperature. Examples of where thermistors are used include a simple electrical thermometer, a circuit to switch on a fan inside your laptop when it overheats, and a central-heating controller that senses the temperature using a thermistor.

The variation of resistance

B–A*
- Increasing the temperature of a metal does not significantly change the number of conduction electrons.
- Thermistors are made from either metal oxides or semiconductors. In these materials, an increase in temperature frees up more electrons from the atoms, and this causes the resistance to decrease.
- LDRs are made from semiconductors that are responsive to light. As the intensity of light falling on an LDR increases, this frees up more electrons from the atoms and this causes the resistance to decrease.

Improve your grade

Kettle calculations

Higher: Noah buys an electric kettle that supplies 2000 W of power. A full kettle of water requires 200 000 J of energy to bring it to boiling point.

a Assuming that the kettle wastes no energy, how long will it take Noah to boil a full kettle of water?

b The current in the heating element produces a heating effect. Describe how this occurs.   *AO2* [3 marks]

Scalar and vector quantities

Scalars and vectors

- Quantities in physics can be divided into two groups:
  - A **scalar** quantity only has size (or magnitude), e.g. distance, mass, volume, temperature, speed and energy.
  - A **vector** quantity has both magnitude and direction, e.g. displacement, velocity, acceleration and force.
- **Displacement** has a size that is equal to the distance from a specified point. It also has direction. For example, in the graph opposite, the aeroplane has travelled a distance of 40 km and has a displacement of 28 km (its size) at a bearing of 45° (its direction) from A.
- **Speed** is defined as the rate of change of distance. Most journeys are not covered at a constant speed. The average speed of a journey can be calculated using the equations:

$$\text{speed (metre per second, m/s)} = \frac{\text{distance (metre, m)}}{\text{time taken (second, s)}} \text{ or } v = \frac{x}{t}$$

- The **velocity** of an object is its speed in a specified direction. It is also defined as the rate of change of displacement.
- The car in the diagram opposite is travelling at a constant speed, but its velocity is changing continuously.

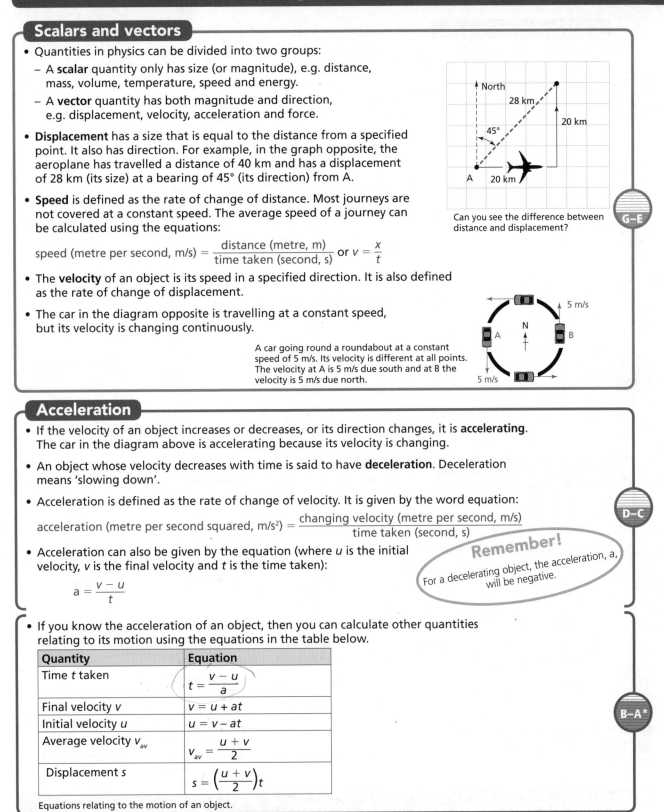

Can you see the difference between distance and displacement?

A car going round a roundabout at a constant speed of 5 m/s. Its velocity is different at all points. The velocity at A is 5 m/s due south and at B the velocity is 5 m/s due north.

G–E

Acceleration

- If the velocity of an object increases or decreases, or its direction changes, it is **accelerating**. The car in the diagram above is accelerating because its velocity is changing.
- An object whose velocity decreases with time is said to have **deceleration**. Deceleration means 'slowing down'.
- Acceleration is defined as the rate of change of velocity. It is given by the word equation:

$$\text{acceleration (metre per second squared, m/s}^2) = \frac{\text{changing velocity (metre per second, m/s)}}{\text{time taken (second, s)}}$$

- Acceleration can also be given by the equation (where u is the initial velocity, v is the final velocity and t is the time taken):

$$a = \frac{v - u}{t}$$

Remember!
For a decelerating object, the acceleration, a, will be negative.

D–C

- If you know the acceleration of an object, then you can calculate other quantities relating to its motion using the equations in the table below.

Quantity	Equation
Time t taken	$t = \dfrac{v - u}{a}$
Final velocity v	$v = u + at$
Initial velocity u	$u = v - at$
Average velocity v_{av}	$v_{av} = \dfrac{u + v}{2}$
Displacement s	$s = \left(\dfrac{u + v}{2}\right)t$

Equations relating to the motion of an object.

B–A*

Improve your grade

Understanding acceleration

Foundation: Syamala is training for the 1200-m race by running round a 200-m track six times. She runs at a constant speed.
- Explain why she is accelerating. *AO1* [2 marks]

Distance–time and velocity–time graphs

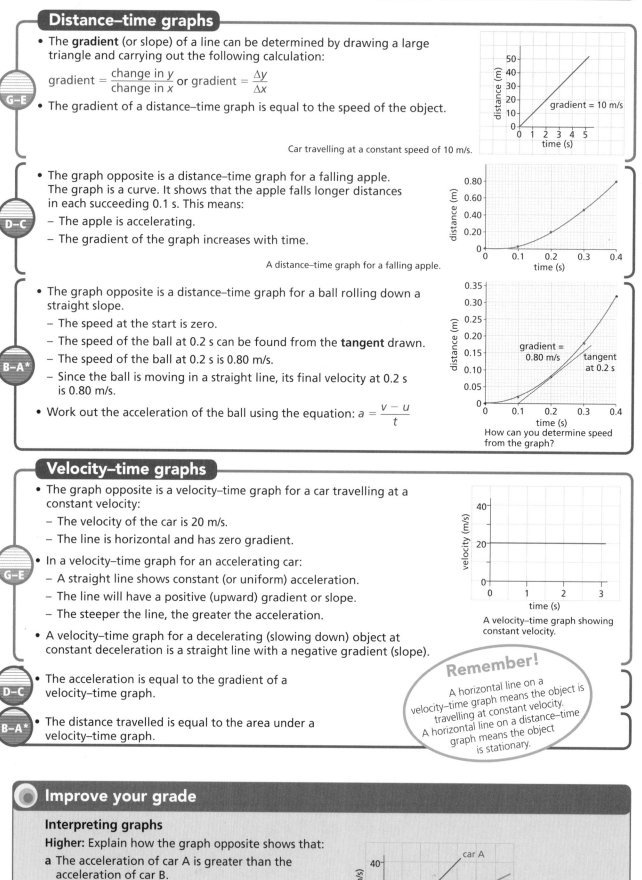

Distance–time graphs

G–E

- The **gradient** (or slope) of a line can be determined by drawing a large triangle and carrying out the following calculation:

$$\text{gradient} = \frac{\text{change in } y}{\text{change in } x} \text{ or gradient} = \frac{\Delta y}{\Delta x}$$

- The gradient of a distance–time graph is equal to the speed of the object.

Car travelling at a constant speed of 10 m/s.

D–C

- The graph opposite is a distance–time graph for a falling apple. The graph is a curve. It shows that the apple falls longer distances in each succeeding 0.1 s. This means:
  - The apple is accelerating.
  - The gradient of the graph increases with time.

A distance–time graph for a falling apple.

B–A*

- The graph opposite is a distance–time graph for a ball rolling down a straight slope.
  - The speed at the start is zero.
  - The speed of the ball at 0.2 s can be found from the **tangent** drawn.
  - The speed of the ball at 0.2 s is 0.80 m/s.
  - Since the ball is moving in a straight line, its final velocity at 0.2 s is 0.80 m/s.
- Work out the acceleration of the ball using the equation: $a = \dfrac{v - u}{t}$

How can you determine speed from the graph?

Velocity–time graphs

G–E

- The graph opposite is a velocity–time graph for a car travelling at a constant velocity:
  - The velocity of the car is 20 m/s.
  - The line is horizontal and has zero gradient.
- In a velocity–time graph for an accelerating car:
  - A straight line shows constant (or uniform) acceleration.
  - The line will have a positive (upward) gradient or slope.
  - The steeper the line, the greater the acceleration.
- A velocity–time graph for a decelerating (slowing down) object at constant deceleration is a straight line with a negative gradient (slope).

A velocity–time graph showing constant velocity.

D–C

- The acceleration is equal to the gradient of a velocity–time graph.

B–A*

- The distance travelled is equal to the area under a velocity–time graph.

Remember!

A horizontal line on a velocity–time graph means the object is travelling at constant velocity. A horizontal line on a distance–time graph means the object is stationary.

◎ Improve your grade

Interpreting graphs

Higher: Explain how the graph opposite shows that:

a The acceleration of car A is greater than the acceleration of car B.

b Car B has travelled further than car A.

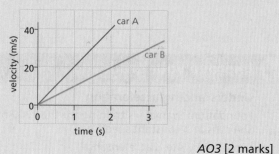

AO3 [2 marks]

Understanding forces

The basics about forces

- A force is a push or a pull exerted by one object on another.
- Force is a **vector** quantity: it has both size (magnitude) and direction.
- Force is measured in units called newtons, N.
- The direction of forces must be taken into account when adding them. For example, in the diagram below, the resultant force of the two 20 N forces can be either 40 N or zero.

  20 N
  20 N

  resultant force = 40 N

  20 N   20 N

  resultant force = 0

  Why are the resultant forces different?

- A **free-body force diagram** shows all the forces acting on an object. Some examples are shown opposite.

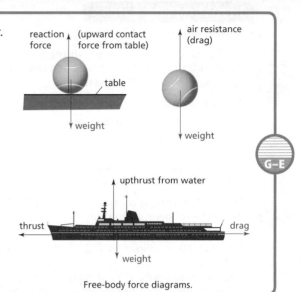

reaction force (upward contact force from table)

air resistance (drag)

table

weight

weight

upthrust from water

thrust   drag

weight

Free-body force diagrams.

Newton's first and third laws

- Sir Isaac Newton's first law of motion states that if there is no resultant force acting on an object, then:
  - if stationary, the object will remain at rest
  - if moving, the object will keep moving at a constant speed in a straight line.
- In the examples of free-body force diagrams above, the forces balance out. Each object is either stationary or moving at a constant speed in a straight line.
- Newton's third law states that when two objects interact, each object exerts an equal but opposite force on the other. We call these equal and opposite forces action and reaction forces.
- A car travelling along a straight road at a constant velocity has the following forces acting on it:
  - the total forward force F between the tyres and the road
  - frictional forces D, including the air resistance or drag opposing the motion of the car
  - the weight W of the car
  - the total upward contact force N provided by the road.
- Since the car is not moving vertically up or down, N is equal to W.
- The resultant force on the car in the horizontal direction is zero. The forces F and D are balanced. The car has no acceleration.

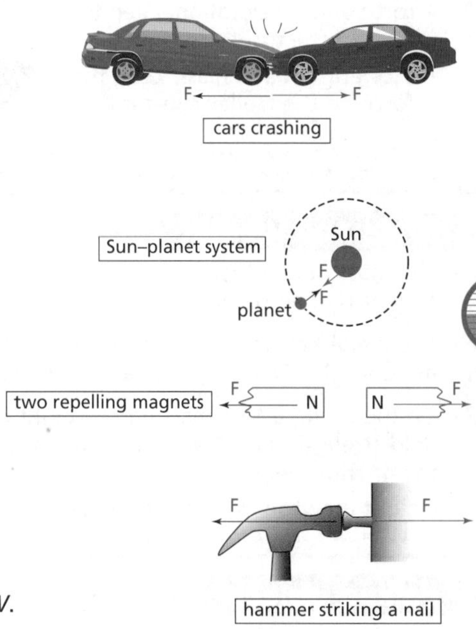

F   F

cars crashing

Sun–planet system   Sun
                         F
                      F
              planet

two repelling magnets   F   F
                         N   N

F   F

hammer striking a nail

Interacting objects. What is special about the size and direction of the force?

Action at a distance

- The gravitational force between the interacting Sun and the Earth acts over a long distance. There is action at a distance.
- According to Newton's third law, the force provided by the Sun on the Earth is equal in size but opposite in direction to the force provided by the Earth on the Sun.
- The gravitational force acting on an object on the Earth is called its **weight**. Every object has weight.
- The Earth interacts with every object on it, including you. You are pulling the Earth towards you with a force equal to your weight.

Improve your grade

Forces and velocity

Foundation: An aeroplane experiences the forces of gravity, upthrust, thrust from the engine and friction from the air.

Explain how it can be travelling at a constant velocity.

AO1 [3 marks]

Force, mass and acceleration

Investigating forces

- If there is a resultant force, then an object will have acceleration.

- Look at the diagram opposite. If the force F on the car is suddenly increased, there will be a horizontal unbalanced force and the car will accelerate.

Remember!
No resultant force means either the object is stationary or it is moving with a constant speed in a straight line.

- The acceleration of the car depends on:
  - the size of the resultant force
  - the mass of the car.

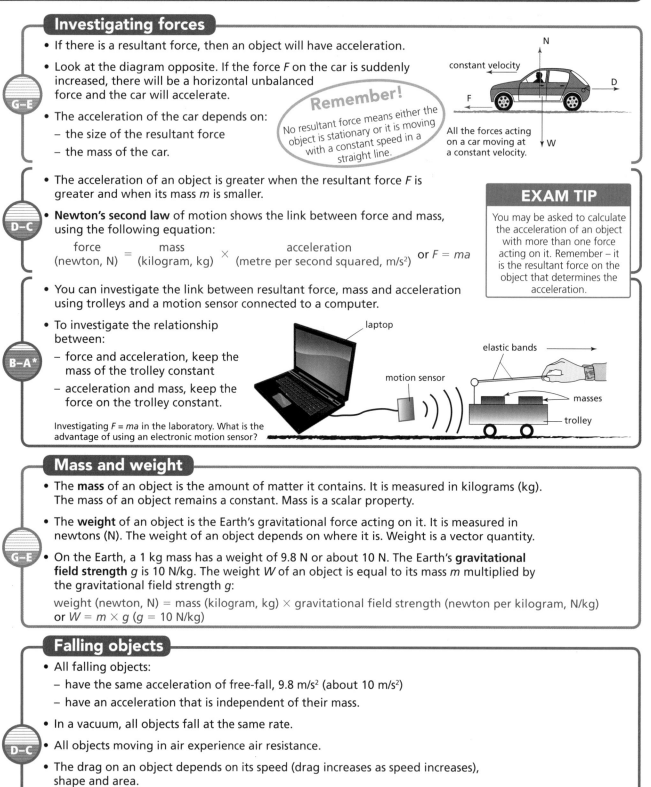

All the forces acting on a car moving at a constant velocity.

- The acceleration of an object is greater when the resultant force F is greater and when its mass m is smaller.

- **Newton's second law** of motion shows the link between force and mass, using the following equation:

$$\frac{\text{force}}{\text{(newton, N)}} = \frac{\text{mass}}{\text{(kilogram, kg)}} \times \frac{\text{acceleration}}{\text{(metre per second squared, m/s}^2\text{)}} \quad \text{or } F = ma$$

EXAM TIP

You may be asked to calculate the acceleration of an object with more than one force acting on it. Remember – it is the resultant force on the object that determines the acceleration.

- You can investigate the link between resultant force, mass and acceleration using trolleys and a motion sensor connected to a computer.

- To investigate the relationship between:
  - force and acceleration, keep the mass of the trolley constant
  - acceleration and mass, keep the force on the trolley constant.

Investigating F = ma in the laboratory. What is the advantage of using an electronic motion sensor?

Mass and weight

- The **mass** of an object is the amount of matter it contains. It is measured in kilograms (kg). The mass of an object remains a constant. Mass is a scalar property.

- The **weight** of an object is the Earth's gravitational force acting on it. It is measured in newtons (N). The weight of an object depends on where it is. Weight is a vector quantity.

- On the Earth, a 1 kg mass has a weight of 9.8 N or about 10 N. The Earth's **gravitational field strength** g is 10 N/kg. The weight W of an object is equal to its mass m multiplied by the gravitational field strength g:

  weight (newton, N) = mass (kilogram, kg) × gravitational field strength (newton per kilogram, N/kg)
  or $W = m \times g$ $(g = 10$ N/kg$)$

Falling objects

- All falling objects:
  - have the same acceleration of free-fall, 9.8 m/s² (about 10 m/s²)
  - have an acceleration that is independent of their mass.

- In a vacuum, all objects fall at the same rate.

- All objects moving in air experience air resistance.

- The drag on an object depends on its speed (drag increases as speed increases), shape and area.

- Air resistance increases as speed increases, until the air resistance is equal to the weight of the object. At this point, the resultant force on the object is zero. The object has reached a constant velocity known as **terminal velocity**.

Improve your grade

Falling objects

Foundation: Kerri is a professional skydiver. When she jumps out of an aeroplane, she initially accelerates, then falls at a steady speed. Explain why.

AO1 [3 marks]

Stopping distance

Thinking, braking and stopping

- Cars rely on friction to stop, using their brakes.
- The **stopping distance** of a car is made up of two parts:
  - **Thinking distance** is the distance travelled by the car as the driver reacts to apply the brakes. It can be calculated using the equation:

  thinking distance (m) = speed of car (m/s) × reaction time (s)

  - **Braking distance** is the distance travelled by the car while the brakes are applied before the car comes to a stop:

  stopping distance = thinking distance + braking distance

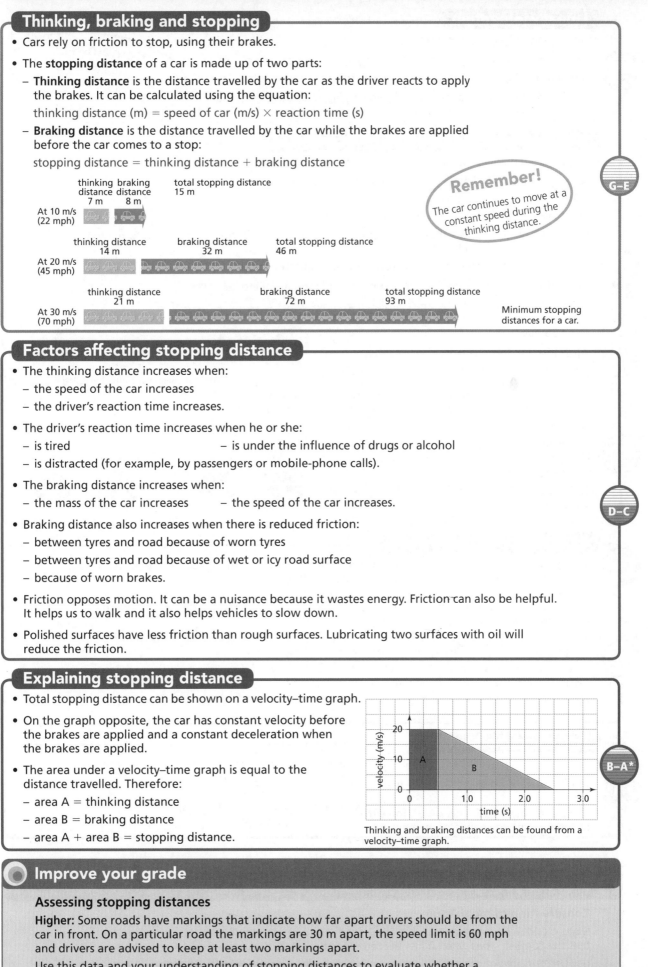

Remember!
The car continues to move at a constant speed during the thinking distance.

At 10 m/s (22 mph): thinking distance 7 m, braking distance 8 m, total stopping distance 15 m

At 20 m/s (45 mph): thinking distance 14 m, braking distance 32 m, total stopping distance 46 m

At 30 m/s (70 mph): thinking distance 21 m, braking distance 72 m, total stopping distance 93 m

Minimum stopping distances for a car.

G–E

Factors affecting stopping distance

- The thinking distance increases when:
  - the speed of the car increases
  - the driver's reaction time increases.
- The driver's reaction time increases when he or she:
  - is tired
  - is distracted (for example, by passengers or mobile-phone calls).
  - is under the influence of drugs or alcohol
- The braking distance increases when:
  - the mass of the car increases
  - the speed of the car increases.
- Braking distance also increases when there is reduced friction:
  - between tyres and road because of worn tyres
  - between tyres and road because of wet or icy road surface
  - because of worn brakes.
- Friction opposes motion. It can be a nuisance because it wastes energy. Friction can also be helpful. It helps us to walk and it also helps vehicles to slow down.
- Polished surfaces have less friction than rough surfaces. Lubricating two surfaces with oil will reduce the friction.

D–C

Explaining stopping distance

- Total stopping distance can be shown on a velocity–time graph.
- On the graph opposite, the car has constant velocity before the brakes are applied and a constant deceleration when the brakes are applied.
- The area under a velocity–time graph is equal to the distance travelled. Therefore:
  - area A = thinking distance
  - area B = braking distance
  - area A + area B = stopping distance.

Thinking and braking distances can be found from a velocity–time graph.

B–A*

Improve your grade

Assessing stopping distances

Higher: Some roads have markings that indicate how far apart drivers should be from the car in front. On a particular road the markings are 30 m apart, the speed limit is 60 mph and drivers are advised to keep at least two markings apart.

Use this data and your understanding of stopping distances to evaluate whether a separation of two markings is safe in good weather.

AO3 [3 marks]

Momentum

Linear momentum

- All moving objects have **momentum**.
- The linear momentum of an object moving in a straight line is defined as:
  momentum (kg m/s) = mass (kg) × velocity (m/s) or momentum = $m \times v$
- A minus sign implies that an object is moving in the opposite direction (momentum is a vector quantity).

> **Remember!**
> The momentum of an object will be big if the object is travelling quickly and it has a large mass.

Collisions and conservation of momentum

D–C

- The diagram below shows a moving ball, X, about to collide with a stationary ball, Y.
- As they collide, each ball exerts an equal but opposite force on the other:
  - The force from ball Y slows down ball X.
  - The force from ball X makes ball Y move.
- When two objects collide, their total momentum remains constant as long as no external forces are acting. This is the **principle of conservation of momentum**:
  total momentum before collision = total momentum after collision
- In the diagram, ball Y gains the same amount of momentum lost by ball X.
- The balls stick together after the collision and move with a common velocity v: $(3.0 \times 2.0) + (1.5 \times 0) = (4.5 \times v)$
  $v = 1.33$ m/s
- After the collision, the balls have a velocity of 1.33 m/s towards the right.

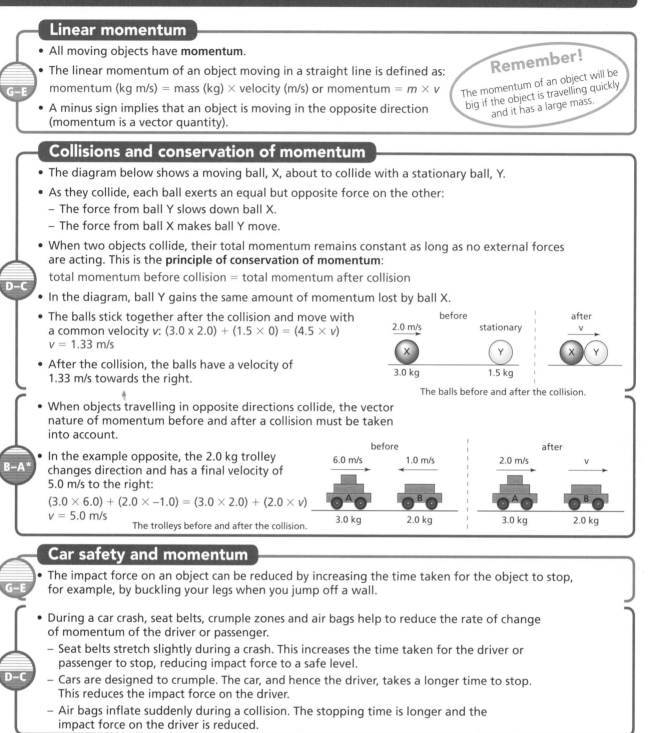

The balls before and after the collision.

B–A*

- When objects travelling in opposite directions collide, the vector nature of momentum before and after a collision must be taken into account.
- In the example opposite, the 2.0 kg trolley changes direction and has a final velocity of 5.0 m/s to the right:
  $(3.0 \times 6.0) + (2.0 \times -1.0) = (3.0 \times 2.0) + (2.0 \times v)$
  $v = 5.0$ m/s

The trolleys before and after the collision.

Car safety and momentum

G–E

- The impact force on an object can be reduced by increasing the time taken for the object to stop, for example, by buckling your legs when you jump off a wall.

D–C

- During a car crash, seat belts, crumple zones and air bags help to reduce the rate of change of momentum of the driver or passenger.
  - Seat belts stretch slightly during a crash. This increases the time taken for the driver or passenger to stop, reducing impact force to a safe level.
  - Cars are designed to crumple. The car, and hence the driver, takes a longer time to stop. This reduces the impact force on the driver.
  - Air bags inflate suddenly during a collision. The stopping time is longer and the impact force on the driver is reduced.

Force and momentum

B–A*

- If a force F acts on an object of mass m for a time t, the velocity of the object changes from u to v.
  force = mass × acceleration $F = ma = m \times \dfrac{(v - u)}{t}$ $F = \dfrac{mv - mu}{t}$
- This can be described in words as: force (N) = $\dfrac{\text{change in momentum (kg m/s)}}{\text{time (s)}}$
  or force = rate of change of momentum

● Improve your grade

Conservation of momentum

Higher: Caitlin likes to play snooker. She is learning to hit a red ball with the cue ball in such a way that the cue ball stops when it hits the red. Both the cue ball and the red ball have the same mass.

Use the principle of conservation of momentum to describe the velocity of the balls before and after the collision.

AO2 [3 marks]

34 **Unit P2**

Topic 4: 4.4, 4.5, 4.6, 4.7, 4.8, 4.9

Work, energy and power

Work and energy

- There is **work done** whenever a force is applied on an object and it moves:

  work done = force (newton, N) × distance moved in the direction of the force (metre, m)
  or $E = F \times d$

- Work done is measured in newton metres or **joules** (J). 1 joule is the work done when a force of 1 newton moves through a distance of 1 metre in the direction of the force.

  work done by a force = energy transferred

- For example, if a force of 40 N is exerted on a box as it moves a distance of 5.0 m along the floor, the work done by the force is calculated:

  $F \times d = 40 \times 5.0 = 200$ J

- In the above example, the work done on the box is transferred to heat between the box and the floor.

- In another example, a person of weight 400 N on an escalator climbs a vertical height of 6.0 m. Here, the work done against the weight is calculated:

  $F \times d = 400 \times 6.0 = 2400$ J

- In this example, the work done is transferred to **gravitational potential energy**.

G–E

Power

- **Power** is the rate at which work is done, or the rate at which energy is transferred.

- Power is measured in watts (W). 1 watt = 1 joule per second (J/s).

- Power is also measured in kW (1000 W) and MW (1 000 000 W).

- Power can be calculated using the equation:

  $$\text{power} = \frac{\text{work done (joule, J)}}{\text{time taken (second, s)}}$$

  or (where E is the work done and t is the time taken):

  $$P = \frac{E}{t}$$

- For example, if a crane lifts a weight of 1200 N through a vertical height of 40 m in 5 minutes (300 s), the power can be calculated:

  work done E by the crane = $F \times d = 1200 \times 40 = 48\,000$ J

  power $P = \dfrac{E}{t} = \dfrac{48\,000}{300} = 160$ W

D–C

Power at a constant speed

- The diagram below shows a car travelling at a constant speed v. The forward force provided by the car is equal to the drag F.

- All the work done by the car engine is transferred into heat.

- The output power of the car can be calculated:

  $$\text{output power} = \frac{\text{work done}}{\text{time}} = \frac{F \times d}{t} = F \times \left(\frac{d}{t}\right)$$

  speed $v = \left(\dfrac{d}{t}\right)$ therefore   output power $P = Fv$

- The equation $P = Fv$ will only work if the car is travelling at a constant velocity.

B–A*

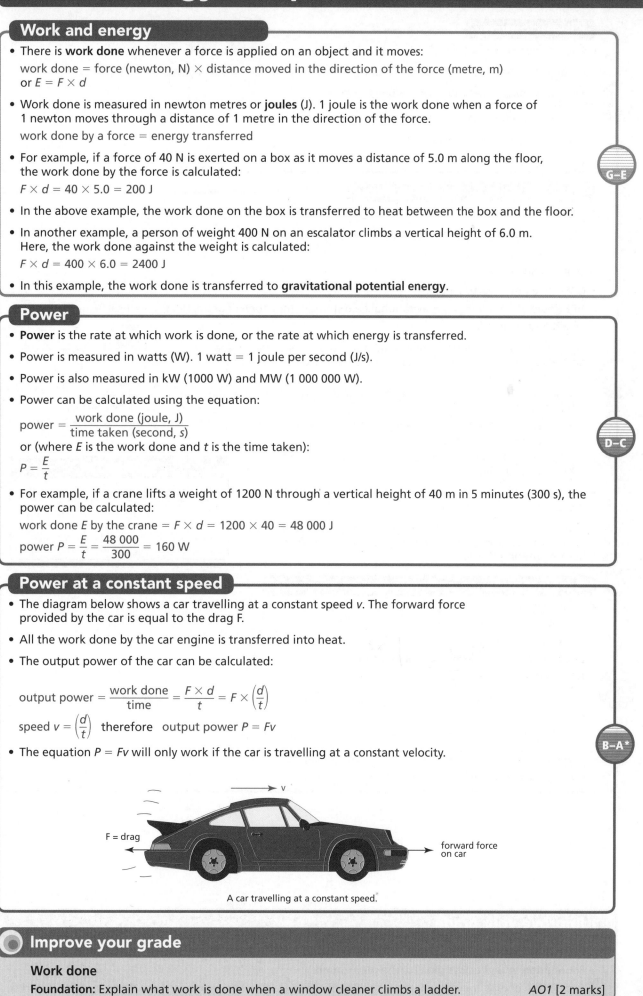

A car travelling at a constant speed.

Improve your grade

Work done

Foundation: Explain what work is done when a window cleaner climbs a ladder. *AO1* [2 marks]

KE, GPE and conservation of energy

Kinetic energy

G–E

- **Kinetic energy** (KE) is energy of movement. All moving objects have kinetic energy.
- An object's kinetic energy is determined by its mass and speed (or velocity), using the equation:

  KE = ½ × mass × velocity²   or $KE = \frac{1}{2} \times m \times v^2$
- KE is measured in joules (J).
- Mass is measured in kilograms (kg).
- Velocity is measured in metres per second (m/s).

Gravitational potential energy

D–C

- The diagram below right shows an object of mass m moved from A to B through a vertical height h.
- The object gains **gravitational potential energy** (GPE). Potential energy is stored energy.
- GPE is calculated:

  GPE = work done

  GPE = weight × vertical height

  GPE (joule, J) = mass (kg) × gravitational field strength (newton/kilogram, N/kg) × vertical height (metre, m)

  or GPE = $m \times g \times h$
- The principle of conservation of energy states that energy can neither be created nor destroyed, it can only be transformed into different forms. For example:
  - When a cyclist brakes, kinetic energy is transformed into heat in the brakes and sound.
  - In a petrol car, the **chemical energy** of the fuel is changed into kinetic energy, heat and sound.
  - When a parachutist falls, his or her GPE is transformed into KE, heat and a bit of sound.
- At the top of a diving board, a diver has GPE. As he or she falls, the GPE is converted into KE.
- In the previous example, the principle of conservation of energy can be used to determine the speed v of the diver just before entering the pool.

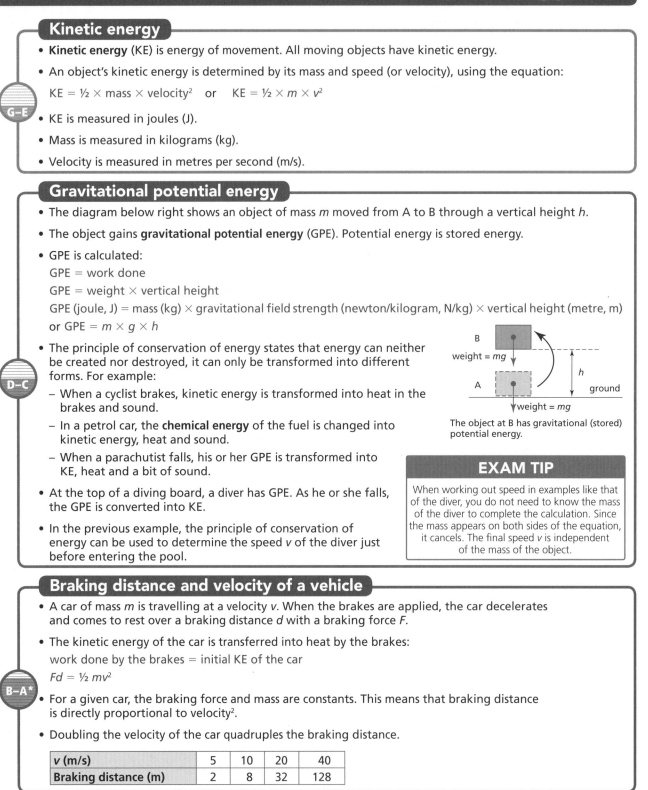

weight = mg

B

A

h

ground

weight = mg

The object at B has gravitational (stored) potential energy.

EXAM TIP

When working out speed in examples like that of the diver, you do not need to know the mass of the diver to complete the calculation. Since the mass appears on both sides of the equation, it cancels. The final speed v is independent of the mass of the object.

Braking distance and velocity of a vehicle

B–A*

- A car of mass m is travelling at a velocity v. When the brakes are applied, the car decelerates and comes to rest over a braking distance d with a braking force F.
- The kinetic energy of the car is transferred into heat by the brakes:

  work done by the brakes = initial KE of the car

  $Fd = \frac{1}{2}mv^2$
- For a given car, the braking force and mass are constants. This means that braking distance is directly proportional to velocity².
- Doubling the velocity of the car quadruples the braking distance.

v (m/s)	5	10	20	40
Braking distance (m)	2	8	32	128

Improve your grade

Energy transfers

Foundation: Abeni enjoys slides at the playground.

Explain the energy transfers that occur as Abeni climbs up the slide and then slides down it. *AO2* [3 marks]

Atomic nuclei and radioactivity

Atoms and ions

G–E

- The nucleus of an atom contains neutrons and protons, together referred to as **nucleons**. Around the nucleus are electrons.
- A neutral atom has the same number of electrons and protons.
- An ion is a charged atom that has lost or gained electrons:
  - Adding electrons makes an atom a negative ion.
  - Removing electrons makes an atom a positive ion.
- Positive ions can be created by:
  - rubbing insulators together (the friction removes electrons from the atoms of one insulator)
  - heating a gas (thermal energy ionises the gas atoms; electrons of the atoms gain energy and fly off).

D–C

- The nucleus of an atom is represented as: $^A_Z X$
- X is the chemical symbol for the element.
- A is the **nucleon number** (or **mass number**), which is the total number of neutrons and protons in the nucleus.
- Z is the **proton number** (or **atomic number**) – the number of protons in the nucleus.
- The number of neutrons N in the nucleus is equal to (A – Z).

Isotopes

B–A*

- **Isotopes** of an element are nuclei that have the same number of protons but a different number of neutrons.
- The isotopes of a particular element have the same chemical properties because they all have the same number of electrons.

Radioactivity

G–E

- Some isotopes are unstable. Over time, the nucleus breaks up and emits a particle or wave in an attempt to become more stable. This is called radioactive decay.
- There are three types of nuclear radiations:
  - **Alpha (α) particles**: each alpha particle is identical to a helium nucleus, with two protons and two neutrons.
  - **Beta (β) particles**: each beta particle is an electron emitted from inside the nucleus.
  - **Gamma (γ) rays**: gamma rays are electromagnetic waves of very short wavelength.
- The radiations from radioactive materials carry energy and can cause **ionisation**.
- Ionisation is the process of removing electrons from atoms. This leaves behind positive ions.

- The table below summarises the main properties of different radiations.

D–C

Radiation	What is it?	Charge	Typical speed (m/s)	Mass	Ionising effect	Penetration
α	Helium nucleus ($^4_2$He)	+2	10 million	4	Strong	Stopped by paper, skin or about 6 cm of air
β	Electron ($^0_{-1}$e)	−1	100 million	0.00055	Weak	Stopped by a few millimetres of aluminium
γ	Short-wavelength electromagnetic wave	0	3.0×10^8	0	Very weak	Never completely stopped, but reduced significantly by thick lead or concrete

B–A*

- Before it decays, a nucleus is known as the 'parent'. The nucleus left behind after decay is known as the 'daughter'.
- In alpha decay, two protons and two neutrons are removed from the nucleus. The proton number decreases by two and the nucleon number decreases by four.

Remember!
In radioactive decay, the daughter nucleus is of a *different* element.

Improve your grade

Radioactive decay

Higher: Alice has learnt at school that radioactive decay is random and cannot be predicted. Andrew argues that it must be possible to predict when radiation will occur, otherwise it would not be useable in the treatment and diagnosis of medical conditions.

Who is correct?

AO2 [2 marks]

Nuclear fission

How fission works

- Nuclear reactions produce energy:
  - In radioactive decay, the kinetic energy of the alpha or beta particles emitted from the nuclei can be used to generate electricity on a small scale.
  - **Fusion reactions** cause the energy generated by the Sun and the stars.
  - **Fission reactions** take place in a nuclear reactor to generate electricity on a large scale.

G–E

- Fission means 'splitting' the nucleus. In a fission reaction:
  - A slow-moving neutron is absorbed by a $^{235}_{92}U$ nucleus, creating an unstable $^{235}_{92}U$ nucleus.
  - The new nucleus splits into two smaller nuclei (**daughter nuclei**) and two or more fast-moving neutrons.
  - Energy is released as the kinetic energy of the daughter nuclei and the neutrons.

A fission reaction of the uranium-235 nucleus by a slow-moving neutron.

D–C

- Radioactive decay happens naturally – we cannot control it. However, scientists can trigger fission reactions by bombarding uranium-235 nuclei with neutrons.

- In a large sample of uranium, the fast-moving neutrons from the fission reactions can go on to split other uranium-235 nuclei. This is called a **chain reaction**.

Chain reaction.

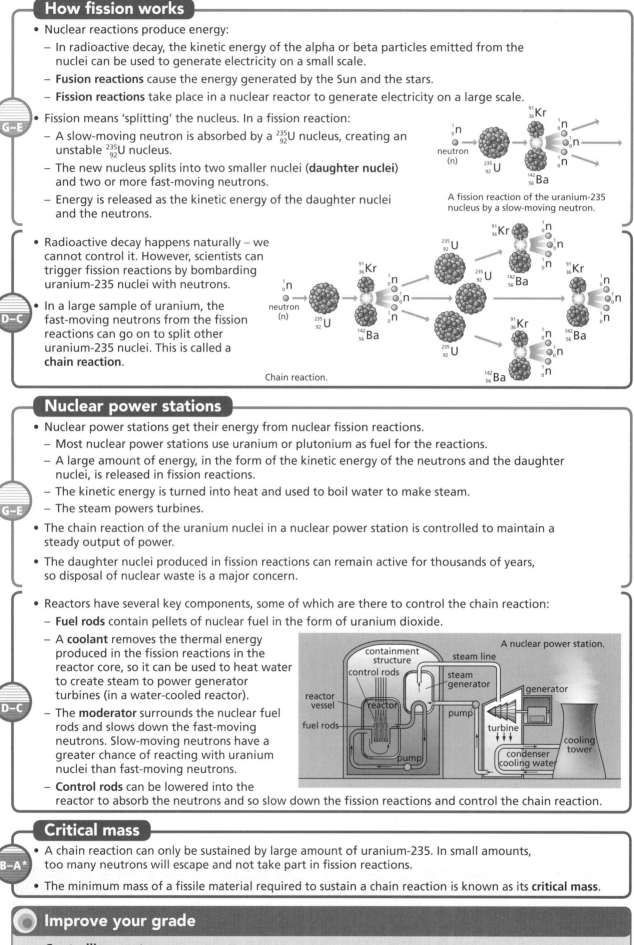

Nuclear power stations

- Nuclear power stations get their energy from nuclear fission reactions.
  - Most nuclear power stations use uranium or plutonium as fuel for the reactions.
  - A large amount of energy, in the form of the kinetic energy of the neutrons and the daughter nuclei, is released in fission reactions.
  - The kinetic energy is turned into heat and used to boil water to make steam.
  - The steam powers turbines.

G–E

- The chain reaction of the uranium nuclei in a nuclear power station is controlled to maintain a steady output of power.

- The daughter nuclei produced in fission reactions can remain active for thousands of years, so disposal of nuclear waste is a major concern.

- Reactors have several key components, some of which are there to control the chain reaction:
  - **Fuel rods** contain pellets of nuclear fuel in the form of uranium dioxide.

D–C

  - A **coolant** removes the thermal energy produced in the fission reactions in the reactor core, so it can be used to heat water to create steam to power generator turbines (in a water-cooled reactor).

A nuclear power station.

containment structure
control rods
steam line
steam generator
generator
reactor vessel
reactor
pump
fuel rods
turbine
cooling tower
pump
condenser
cooling water

  - The **moderator** surrounds the nuclear fuel rods and slows down the fast-moving neutrons. Slow-moving neutrons have a greater chance of reacting with uranium nuclei than fast-moving neutrons.
  - **Control rods** can be lowered into the reactor to absorb the neutrons and so slow down the fission reactions and control the chain reaction.

Critical mass

B–A*

- A chain reaction can only be sustained by large amount of uranium-235. In small amounts, too many neutrons will escape and not take part in fission reactions.

- The minimum mass of a fissile material required to sustain a chain reaction is known as its **critical mass**.

Improve your grade

Controlling neutrons

Foundation: Explain how the neutrons in a nuclear power station are controlled.

AO1 [2 marks]

Fusion on the Earth

Fission and fusion

- In a nuclear fission reaction, a neutron is used to split a uranium-235 nucleus to produce two radioactive daughter nuclei and two or more neutrons.

- The energy released in fission reactions is used by nuclear reactors to produce electricity.

- Nuclear fusion is a nuclear reaction in which two smaller, lighter nuclei join or fuse together to produce one larger nucleus. Fusion reactions produce a vast amount of energy.

- Extremely high temperatures are required for fusion reactions to take place.

- The energy source that keeps our Sun and other stars burning is the fusion of hydrogen and other lighter nuclei.

- The bottom diagram opposite shows the fusion of two isotopes of hydrogen, deuterium and tritium. The reaction produces a stable nucleus of helium and a neutron.

Remember!
Nuclear fission is the splitting of nuclei, nuclear fusion is the joining together of nuclei. Both reactions produce large amounts of energy.

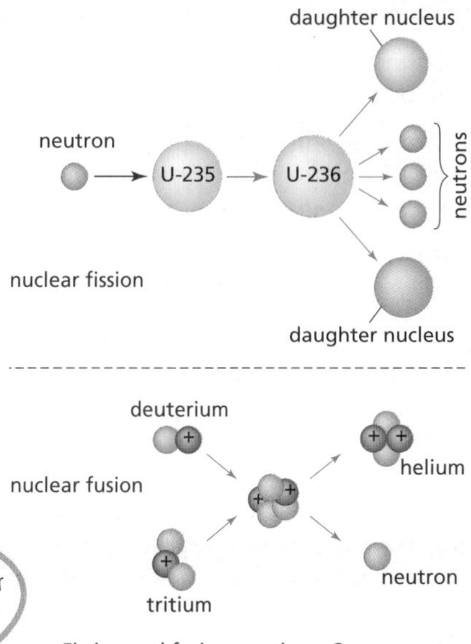

Fission and fusion reactions. Can you spot any differences?

G–E

Cold fusion

- In 1989, scientists Stanley Pons and Martin Fleischmann announced that they had produced nuclear fusion at room temperature. This theory became known as **cold fusion**. They claimed their experiment had produced large amounts of thermal energy.

- The announcement gained worldwide publicity. However, Pons and Fleischmann were criticised by many scientists because they had not published enough technical details of their experiment for other scientists to reproduce their results.

- The majority of scientists now reject Pons and Fleischmann's theory of cold fusion because their theory could not be validated by reproducing their experiment.

D–C

Bringing fusion to the Earth

- Fusion reactions are more difficult to trigger than fission reactions because hydrogen nuclei (protons) are positively charged, and therefore repel each other.

- Increasing the speed at which the nuclei move improves the chances of a fusion reaction taking place.

- At temperatures around 10 million °C, hydrogen nuclei move rapidly enough to overcome the electrostatic repulsive forces and join together in fusion reactions. Nuclear fusion cannot take place at low temperatures and pressures.

- In order to create fusion, hydrogen nuclei must be heated to temperatures of about 100 million °C and contained by very strong magnetic fields produced by super-cooled electromagnets. It is difficult to create these conditions on Earth.

Remember!
Particles move faster at higher temperatures.

B–A*

How science works

You should be able to:

- analyse the reasons that Pons and Fleischmann (and scientists like them) were criticised by the wider scientific community

- understand why the theory of cold fusion has been largely rejected

- explain the reasons why it is necessary for other scientists to be able to reproduce an experiment like theirs.

Improve your grade

The future of fusion

Foundation: Scientists hope that fusion reactors will soon be able to produce energy for human consumption.

a What fuel will be used in these reactors?

b Explain how nuclear fusion occurs.   *AO1/AO2* [3 marks]

Background radiation

Radioactive rocks

- Everything around us is slightly radioactive. This is due to **background radiation**.

- Background radiation comes from a variety of sources and can be detected by a Geiger counter.

- Rocks are naturally radioactive because they contain small traces of radioactive isotopes. Granite is slightly more radioactive than other rocks because it contains higher levels of uranium atoms.

- Uranium nuclei decay naturally over time to produce **radon** nuclei. Radon is a colourless and odourless radioactive gas.

- Houses built over granite can trap radon gas.

- Exposure to radioactive radon can lead to lung cancer.

- Different areas of the UK have different levels of background radiation due to varying amounts of radioactive sources in that region.

Remember!
Background radiation is random, low-level radiation that is present everywhere on Earth.

G–E

Origins of background radiation

- Most background radiation comes from natural sources. These include:
  - *Cosmic rays:* energetic particles such as electrons, protons and neutrinos that come from the Sun and outer space. They penetrate the Earth's atmosphere to reach the surface. The danger from cosmic rays increases with altitude because there is less atmosphere to stop the radiation.
  - *Rocks:* rocks such as granite contain uranium, which decays to produce radioactive radon gas.
  - *Food:* all foods will have minute traces of radioactive nuclei.

- Humans have also had a small effect on background radiation, through:
  - nuclear power stations
  - fallout from previous nuclear weapons tests, explosions and accidents
  - radiation from equipment or waste from hospitals and industry.

D–C

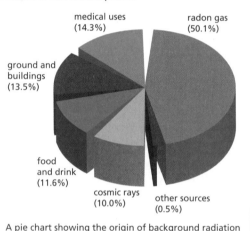

medical uses (14.3%)
radon gas (50.1%)
ground and buildings (13.5%)
food and drink (11.6%)
cosmic rays (10.0%)
other sources (0.5%)

A pie chart showing the origin of background radiation for an average individual in the UK.

Dangers of radon gas

- Radon gas rises from ground that contains granite. Radon gas is particularly dangerous if it remains trapped in the walls of buildings or under floorboards.

- Radon-222 is one of the isotopes of radon gas. It is a non-reactive **noble gas** and itself is not a health hazard.

- Radon-222 nuclei are produced by the decay of radium-226 nuclei.

- The nuclei decay by alpha emission with a short half-life of 3.8 days.

- The daughter nuclei of radon-222 also emit alpha particles.

- It is believed that radon-222 may cause cellular damage in the lungs.

B–A*

Improve your grade

Differences in background radiation

Foundation: Joe and Fred are pen-pals who live in different parts of the UK. They have both been learning about background radiation at school and have discovered that the level of background radiation where they live is different.

Explain why this is.

AO1 [3 marks]

Uses of radioactivity

Industrial uses of radioactivity

- The three types of ionising radiation – alpha particles, beta particles and gamma rays – have practical uses in a number of industries.

- Irradiating food with gamma rays prolongs its shelf-life. The gamma rays kill off **microorganisms** on the food even after it has been packaged.

- In the metal industry, gamma rays are used to check the quality of welding or to detect cracks in metals.

- Beta particles are used in the water industry to detect leaks in underground pipes. A beta-particle emitting radioactive material, known as a **tracer**, is fed into the pipe. Above ground, a radiation detector detects increased levels of radiation.

Remember! Radioactive materials emit alpha particles, beta particles and gamma rays.

G–E

- The thickness of paper can be controlled using a source of beta particles (see diagram).

- Pressure is applied by the rollers to control the thickness of the paper. A beta source is placed above the paper and a radiation detector is placed directly below the source and the paper.

- If the paper is thicker than required, the detector shows an increase in the number of beta particles recorded per unit time. A signal is then sent to the rollers to increase the pressure and reduce the thickness.

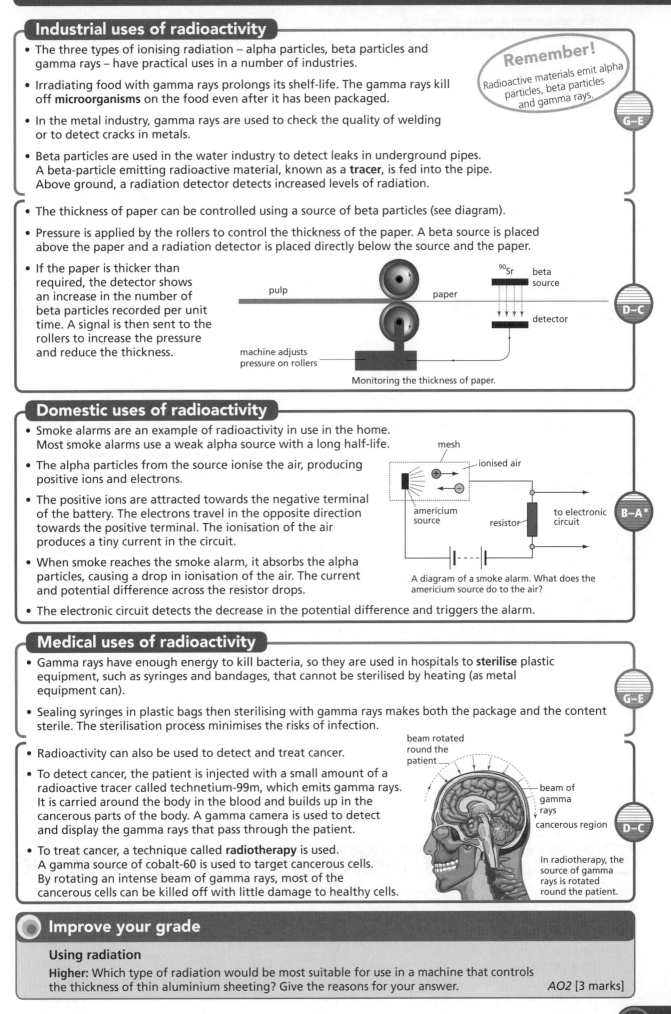

Monitoring the thickness of paper.

D–C

Domestic uses of radioactivity

- Smoke alarms are an example of radioactivity in use in the home. Most smoke alarms use a weak alpha source with a long half-life.

- The alpha particles from the source ionise the air, producing positive ions and electrons.

- The positive ions are attracted towards the negative terminal of the battery. The electrons travel in the opposite direction towards the positive terminal. The ionisation of the air produces a tiny current in the circuit.

- When smoke reaches the smoke alarm, it absorbs the alpha particles, causing a drop in ionisation of the air. The current and potential difference across the resistor drops.

- The electronic circuit detects the decrease in the potential difference and triggers the alarm.

A diagram of a smoke alarm. What does the americium source do to the air?

B–A*

Medical uses of radioactivity

- Gamma rays have enough energy to kill bacteria, so they are used in hospitals to **sterilise** plastic equipment, such as syringes and bandages, that cannot be sterilised by heating (as metal equipment can).

- Sealing syringes in plastic bags then sterilising with gamma rays makes both the package and the content sterile. The sterilisation process minimises the risks of infection.

G–E

- Radioactivity can also be used to detect and treat cancer.

- To detect cancer, the patient is injected with a small amount of a radioactive tracer called technetium-99m, which emits gamma rays. It is carried around the body in the blood and builds up in the cancerous parts of the body. A gamma camera is used to detect and display the gamma rays that pass through the patient.

- To treat cancer, a technique called **radiotherapy** is used. A gamma source of cobalt-60 is used to target cancerous cells. By rotating an intense beam of gamma rays, most of the cancerous cells can be killed off with little damage to healthy cells.

In radiotherapy, the source of gamma rays is rotated round the patient.

D–C

Improve your grade

Using radiation

Higher: Which type of radiation would be most suitable for use in a machine that controls the thickness of thin aluminium sheeting? Give the reasons for your answer.

AO2 [3 marks]

Activity and half-life

Radioactivity and half-life

G–E
- Radioactive decay is random and spontaneous, which means it cannot be predicted and that it is not affected by external conditions.
- Some nuclei are very unstable and decay very quickly. Others take a long time to decay.
- The **half-life** of an isotope is the average time it takes for half of the undecayed nuclei in a sample to decay. Half-life can be micro-seconds or thousands of years.

D–C
- The rate of decay of a source's nuclei is known as its **activity**.
- Activity is measured in **becquerel** (Bq) – 100 Bq means that 100 nuclei decay per second and that 100 alpha or beta particles are emitted per second.
- Activity is directly proportional to the number of undecayed nuclei in a source, i.e. activity doubles when the number of nuclei is doubled.
- Activity is inversely proportional to the half-life of the isotope.
- As radioactive nuclei decay, there are fewer undecayed nuclei. The activity of a source thus decreases over time.

> **Remember!**
> A source with a short half-life will have a large activity.

Exponential decay

B–A*
- A sample with a half-life of 15 hours will have half the original number of nuclei after 15 hours, a quarter of the original number of nuclei after 30 hours, and so on.
- The activity will also reduce to half its original value after 15 hours, to a quarter of its original value after 30 hours etc.
- This type of decay is known as **exponential decay**.

The activity of a source decreases with time. What happens to the activity after one half-life?

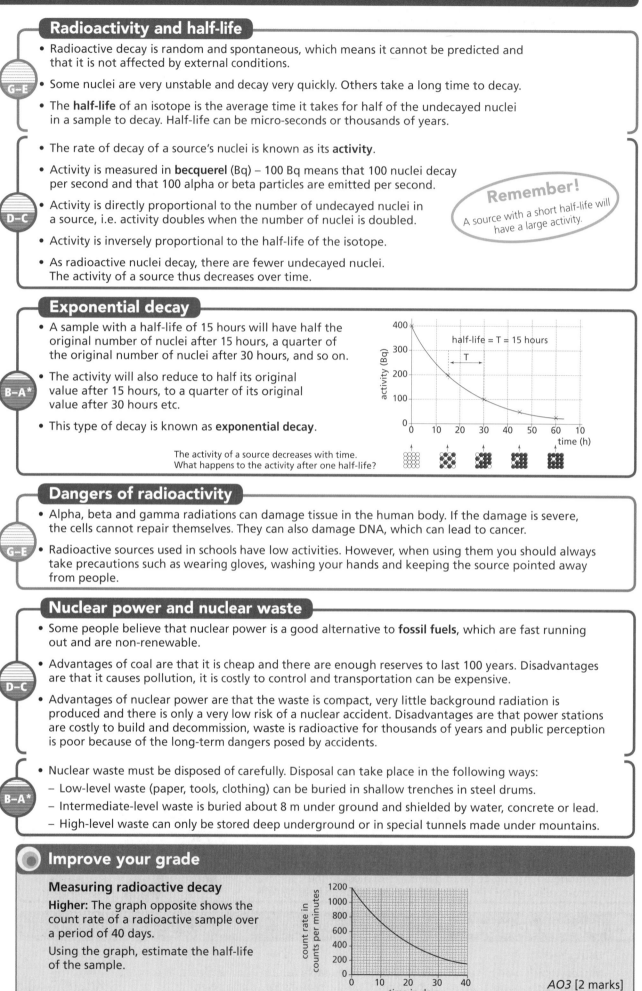

Dangers of radioactivity

G–E
- Alpha, beta and gamma radiations can damage tissue in the human body. If the damage is severe, the cells cannot repair themselves. They can also damage DNA, which can lead to cancer.
- Radioactive sources used in schools have low activities. However, when using them you should always take precautions such as wearing gloves, washing your hands and keeping the source pointed away from people.

Nuclear power and nuclear waste

D–C
- Some people believe that nuclear power is a good alternative to **fossil fuels**, which are fast running out and are non-renewable.
- Advantages of coal are that it is cheap and there are enough reserves to last 100 years. Disadvantages are that it causes pollution, it is costly to control and transportation can be expensive.
- Advantages of nuclear power are that the waste is compact, very little background radiation is produced and there is only a very low risk of a nuclear accident. Disadvantages are that power stations are costly to build and decommission, waste is radioactive for thousands of years and public perception is poor because of the long-term dangers posed by accidents.

B–A*
- Nuclear waste must be disposed of carefully. Disposal can take place in the following ways:
  - Low-level waste (paper, tools, clothing) can be buried in shallow trenches in steel drums.
  - Intermediate-level waste is buried about 8 m under ground and shielded by water, concrete or lead.
  - High-level waste can only be stored deep underground or in special tunnels made under mountains.

Improve your grade

Measuring radioactive decay

Higher: The graph opposite shows the count rate of a radioactive sample over a period of 40 days.

Using the graph, estimate the half-life of the sample.

AO3 [2 marks]

Topic 6: 6.4, 6.5, 6.6, 6.7, 6.8, 6.9, 6.10, 6.11, 6.12

P2 Summary

An insulator can be charged with static electricity by friction, which results in the transfer of electrons. Like charges attract and unlike charges repel.

The dangers of static electricity can be minimised by earthing.

Static and current electricity

Electric current is the rate of flow of charge. Cells and batteries supply direct current (d.c.), which is the movement of charge in one direction only.

A variable resistor can be used to change the resistance in a circuit.

The potential difference across a component can be measured using a voltmeter connected in parallel with the component.

The current can be measured using an ammeter connected in series.

Controlling and using electric current

Electrical power = current × potential difference

Energy transferred = current × potential difference × time

Motion and forces

If the resultant force on a body is zero it will remain at rest or continue to move at the same velocity.

If the resultant force acting on a body is not zero, it will accelerate in the direction of the resultant force according to the equation force = mass × velocity.

Displacement, velocity, acceleration and force are vector quantities. They all take into account direction as well as magnitude.

Acceleration = change in velocity / time

Acceleration can be determined from the gradient of a velocity–time graph.

The stopping distance of a vehicle is made up of the thinking distance and the braking distance.

Momentum, energy, work and power

Momentum = mass × velocity

The rate of change of momentum can be used to explain how safety features of vehicles, such as air bags and crumple zones, reduce the force.

Work done = force × distance moved in the direction of the force

Power is the rate of doing work.

Gravitational potential energy = mass × gravitational field strength × vertical height.

Kinetic energy = half mass × velocity squared

Energy is conserved in energy transfers.

An alpha particle is equivalent to a helium nucleus, a beta particle is an electron emitted from the nucleus and a gamma ray is electromagnetic radiation.

Nuclear fission and nuclear fusion

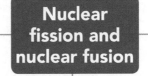

The fission of U-235 produces two daughter nuclei and two or more neutrons.

The chain reaction is controlled in a nuclear reactor by the action of moderators and control rods.

Heat energy from the chain reaction is converted into electrical energy in a nuclear power station.

Nuclear fusion is the creation of larger nuclei from smaller nuclei, accompanied by a release of energy, and is the energy source for stars.

Radioactivity has many uses, including in smoke alarms, sterilisation of equipment, and tracing and gauging thickness. It can also be used to diagnose and treat cancer.

Advantages and disadvantages of using radioactive materials

Issues associated with nuclear power include the lack of carbon dioxide emissions, risks, public perception, waste disposal and safety.

The activity of a radioactive source decreases over time. The half-life of a radioactive isotope is the time taken for half the undecayed nuclei to decay.

Intensity of radiation

Radiation

G–E

- Radiation (in physics) means any form of energy which can travel through space as a wave or a particle.
- Some radiations ionise the atoms of the material they pass through. These radiations are known as ionising radiations.
- Non-ionising radiations include:
  - Radio waves from a transmitter
  - Microwaves from a mobile phone
  - Infrared radiation from a heater
  - Light from a lamp or laser
  - Ultrasound radiation used in hospitals for ultrasound scans.
- Ionising radiations include:
  - Ultraviolet radiation from the Sun
  - X-rays from an X-ray machine
  - Gamma radiation from a radioactive isotope
  - Alpha or beta particles from radioactive isotopes.
- Radiation travels away from a source in all directions.

Intensity of radiation

D–C

- The intensity of radiation is defined as the power or the radiation per unit area, and it has the unit Watts per metre squared (W/m^2).
- The intensity of radiation at a point a distance away from a source:
  - decreases with distance from the source
  - depends on the material the radiation is travelling through.

Remember!
All electromagnetic radiation can travel through a vacuum, but sound waves cannot.

- Intensity of radiation is calculated using the following equation:

  $$Intensity \left(\frac{w}{m^2}\right) = \frac{Power\ of\ incident\ radiation\ (W)}{area\ (m^2)} = \frac{P}{A}$$

- For example: A cycle lamp emits 2 W of light, with a circular beam with diameter 20 cm on the surface of the road. Calculate the intensity of the light:

  $Area\ of\ the\ beam = \pi r^2 = \pi \left(\frac{d}{2}\right)^2 = \pi \left(\frac{0.2}{2}\right)^2$

  $Area = \pi(0.1)^2 = 0.0314\ m^2$

  $Intensity = \frac{Power}{Area} = \frac{2}{0.0314}$

  $Intensity = 63.7\ W/m^2$

- As radiation usually spreads out in all directions from a source, the radiation is spread evenly over the surface area of a sphere ($4\pi r^2$).

B–A*

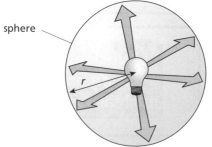

sphere

r

Radiation from a source spreads out uniformly over a sphere.

$$Intensity\ of\ radiation = \frac{Power}{4\pi r^2}$$

- Intensity is inversely proportional to the area.
- Intensity is inversely proportional to the square of the distance.

Improve your grade

Intensity of light

Higher: A street light has a much more powerful bulb than a hand-held battery-powered torch. Explain why the smaller torch can light up the pavement almost as well as the street light.   AO2 [3 marks]

Topic 1: 1.1d, 1.2, 1.3, 1.4

Properties of lenses

Types of lens

- Light is refracted at each surface of a lens.

- A converging lens bends the light towards a focal point. It is fatter in the middle and thinner at the edges.

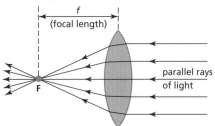

A converging lens. Our eye lens is a converging lens.

- A diverging lens bends the light away from a focal point. It is thicker at the edges and thinner in the middle.

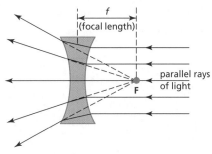

A diverging lends has a virtual principal focus **F**.

- The **focal length** of the lens is the distance between the **principal focus** F and the centre of the lens.

- The fatter the converging lens the shorter its focal length.

Power of a lens

- The **optical power** of a lens is defined as the reciprocal of the focal length.

- *Power of a lens (Dioptre, D)* $= \dfrac{1}{focal\ length\ (m)}$

- The unit of power of a lens is the **Dioptre (D)**.

- The greater the power of a lens the more it refracts or bends the light rays, so a fatter converging lens will be more powerful than a thinner one.

- A converging lens has a real principal focus, and by convention has both a positive power and a positive focal length.

- A diverging lens has a virtual principal focus (the rays of light don't actually meet at the principal focus, they appear to be coming from it), so by convention it has a negative focal length and a negative power.

- Example calculation: A converging lens has a focal length of 20 cm. Calculate the power of this lens:

$$Power = \dfrac{1}{focal\ length\ (m)}$$

$$Power = \dfrac{1}{0.2\ (m)} = +5\ D$$

Remember!

Optical power of a lens has nothing to do with mechanical power or electrical power, which are measured in Watts.

- When two or more lenses are combined, the total power of the combined lens system is the sum of the powers of the two individual lenses.

- The focal length of the lens combination can then be calculated using the equation:

$$focal\ length\ (m) = \dfrac{1}{Power\ (D)}$$

Improve your grade

Power of a lens

Higher: A diverging lens has a focal length of 40 cm. Calculate the optical power of the diverging lens.

AO1 [3 marks]

Lens equation

Ray diagrams for converging lenses

- You can investigate the images formed by lenses using an illuminated object such as a candle flame or a bright filament lamp and a screen. If you set up the object and the lens, you then need to move the screen until you see a focused image of the object.

- Ray diagrams can be used to explain how a converging lens creates an image.

- You need to draw two rays of light from the object (arrow O) through the lens to work out where the image I is formed.

  - Draw a ray of light from the top of object O, parallel to the principal axis of the lens. It will be refracted through the principal focus of the lens.

  - Draw a second ray of light through the centre of the lens. It will not be refracted at all, so will continue in a straight line.

  - The top of the image I will be formed where the two lines cross.

- In this diagram the image is diminished (smaller) and inverted (upside down).

- The distance between the object and the lens is usually given the symbol *u* (object distance), and the distance between the image and the lens is usually given the symbol *v* (image distance).

- A converging lens can be used to produce both magnified and diminished real images, as well as magnified virtual images. The nature of the image depends on the distance between the object and the lens (*u*) and the focal length of the lens (*f*).

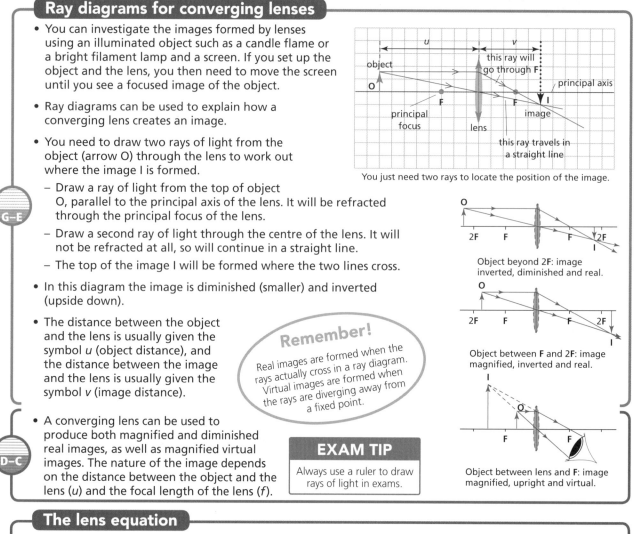

You just need two rays to locate the position of the image.

Object beyond 2F: image inverted, diminished and real.

Object between F and 2F: image magnified, inverted and real.

Object between lens and F: image magnified, upright and virtual.

Remember!
Real images are formed when the rays actually cross in a ray diagram. Virtual images are formed when the rays are diverging away from a fixed point.

EXAM TIP
Always use a ruler to draw rays of light in exams.

The lens equation

- The object distance (*u*), the image distance (*v*), and the focal length (*f*) are related by the lens equation:

$$\frac{1}{\text{focal length (m)}} = \frac{1}{\text{object distance (m)}} + \frac{1}{\text{image distance (m)}}$$

$$\frac{1}{f} = \frac{1}{u} + \frac{1}{v}$$

- There is a 'real is positive' sign convention for use with the lens equation:

Real is positive sign convention

Lens	*f*	*u*	*v*
Converging lens	Positive	Positive	Positive for real image
Diverging lens	Negative	Positive	Negative for virtual image

- For example: A student places a candle flame 20 cm away from a converging lens, and obtains a real focused image of the flame 50 cm away on the other side of the lens. Calculate the focal length of the lens she used.

$$\frac{1}{f} = \frac{1}{u} + \frac{1}{v} = \frac{1}{0.2} + \frac{1}{0.5} = 5 + 2 = 7$$

$$f = \frac{1}{7} = 0.143 \text{ m}$$

How science works

You should be able to:

- investigate changes to images formed by a converging lens with objects at different distances away from the lens in both a quantitative and a qualitative way.

Improve your grade

Images formed using converging lenses

Foundation: The top ray diagram on this page shows how an image is formed by a converging lens. If the object stays in the same place, but you replace the lens with one which has a shorter focal length, explain what will happen to the image. *AO2* (3 marks)

The eye

Structure of the eye

Part of the eye	Description	Function
Cornea	Curved surface at the front of eye	Refracts light towards the lens
Pupil	Hole in the centre of the iris	Allows light into the eye
Iris	Coloured part of the eye	Muscle to change the size of the pupil, to control the amount of light entering the eye.
Lens	Behind the iris	Focuses the light onto the retina
Ciliary muscles	Muscles to change the shape of the lens.	A thin lens is needed to view distant object and a fatter lens needed to view close objects. This is called accommodation.
Retina	Inside surface at back of eye	Contains many light sensitive cells which send messages along the optic nerve to the brain.

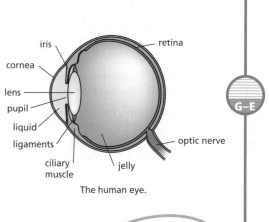

The human eye.

Remember!
Both the cornea and the lens refract light.

- When looking at distant objects the lens needs to be less powerful, so the ciliary muscles relax to make it thinner. The lens is limited so there is a '**far point**', which is the maximum distance the eye can see. In most people the far point is infinity.

- When looking at a close object the lens needs to be more powerful to focus the light onto the retina, so the ciliary muscles tighten to make it fatter. The '**near point**' is the closest distance the eye can focus. This is usually about 25 cm.

Eye defects

Long sight

- Long-sighted people can see distant objects clearly, but cannot focus on close objects.

- The near point for a long-sighted person will be greater than 25 cm, and the image of a closer object will be formed behind the retina.

- This may be caused by the eyeball being too short, the cornea not being curved enough, or the lens not being fat enough.

- Long sight may be corrected by wearing converging lenses.

Correcting long-sightedness.

Short sight

- Short-sighted people can see close objects, but cannot focus on distant objects.

- The far point for short-sighted people is less than infinity, and the image of more distant objects will be in front of the retina.

- This may be caused by the eyeball being too long, the cornea being too curved or the lens being unable to become thin enough.

- Short sight may be corrected by wearing diverging lenses.

- Corrective lenses may be worn as spectacles or contact lenses, but the power of the lenses will not be the same, as the lenses are worn at different distances from the eyes.

Correcting short-sightedness.

- Laser eye surgery can be used as a permanent alternative to wearing glasses or contact lenses.

- Laser eye surgery can be painful for a few days, and like any surgical procedure there is a possibility of infection.

- Most laser eye surgery changes the shape of the cornea, which can correct mild eye defects.

Improve your grade

Short sight

Higher: If a short-sighted person looks at a very distant object, the image is blurred. Explain why.

AO1 [3 marks]

Total internal reflection and endoscopes

Total internal reflection (TIR)

- When light travels from a more optically dense material to a less optically dense material (e.g. glass to air), it is refracted and reflected at the boundary.

- Light travels more slowly in glass than it does in air, so the angle of refraction is larger than the angle of incidence.

- As the angle of incidence increases, the angle of refraction also increases.

- When the angle of incidence equals the **critical angle**, some light is refracted at 90° to the normal (i.e. goes along the glass air boundary) and some light is internally reflected.

- If the angle of incidence is increased further there will be **total internal reflection (TIR)**. Angle of incidence = angle of reflection. There is no refraction.

- The critical angle depends on the type of material.

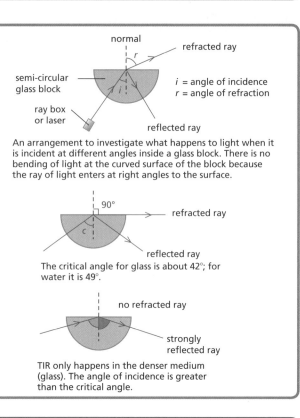

An arrangement to investigate what happens to light when it is incident at different angles inside a glass block. There is no bending of light at the curved surface of the block because the ray of light enters at right angles to the surface.

The critical angle for glass is about 42°; for water it is 49°.

TIR only happens in the denser medium (glass). The angle of incidence is greater than the critical angle.

Optical fibres and endoscopes

- **Optical fibres** are thin, flexible rods of transparent material such as glass.

- Light is transmitted along optical fibres by being totally internally reflected all the way along. Light cannot escape as it always hits the boundary at an angle greater than the critical angle.

- Optical fibres are used for transmitting high-speed signals in broadband networks and computers.

- An **endoscope** is a medical instrument used to see inside the human body. It uses optical fibres, and can be equipped with other surgical instruments to carry out keyhole surgery.

- An endoscope uses two separate bundles of optical fibres: one to illuminate the inside of the patient and one to reflect the image back to an eyepiece or camera.

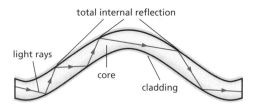

Modern optical fibres have a glass core surrounded by a special coating (known as cladding) and have critical angles as high as 80°.

Snell's Law

- **Snell's Law** is a mathematical relationship between the angles of incidence and refraction and the refractive index of a material for light entering a more dense material.

$$refractive\ index = \frac{\sin i}{\sin r} = \frac{\sin (angle\ outside\ material)}{\sin (angle\ inside\ material)}$$

- As the angle inside the material increases, there comes a point where the angle outside becomes 90°. This inside angle is known as the critical angle.

- At the critical angle (C) the angle on the outside of the material is 90°; and sin 90° = 1

$$refractive\ index\ (n) = \frac{\sin 90}{\sin c} = \frac{1}{\sin c}$$

$$\sin C = \frac{1}{n}$$

- For example, if the refractive index of polycarbonate is 1.6, calculate the critical angle

$$1.6 = \frac{1}{\sin c}$$

$$\frac{1}{1.6} = 0.625 = \sin C$$

$$\sin^{-1} 0.625 = C\ so\ C = 39.2°$$

Improve your grade

Total internal reflection

Foundation: Explain the difference between internal reflection and total internal reflection.   AO1 [3 marks]

Medical uses of ultrasound

Ultrasound diagnoses

- Ultrasound is very high frequency sound waves which cannot be heard by the human ear. The frequency is greater than 20 kHz.

- In hospitals ultrasound of frequency 1.5 MHz is used to form images of the insides of our bodies. The higher the frequency the shorter the wavelength and the smaller the detail of the image can be.

- Ultrasound is generated and received using piezoelectric **transducers**.

- Ultrasound scanners are often linked to computers so a 3-dimensional image can be created.

- Ultrasound scans are safer than X-rays because ultrasound is a non-ionising radiation.

- Ultrasound scans can be used to diagnose:
  - Cysts
  - Tumours
  - Blocked arteries
  - Kidney stones
  - Foetal abnormalities.

- Ultrasound inside our bodies will be reflected and refracted at boundaries between different types of tissue.

- During an ultrasound scan:
  - A piezoelectric source sends out an ultrasonic pulse.
  - As the ultrasound passes from one tissue to another, some is reflected back to a detector.
  - The time between sending out the pulse and receiving the reflected pulse is recorded as *t*. This is the time the pulse takes to travel twice the distance between the source and the tissue boundary.

  $$Distance = \frac{speed\ of\ ultrasound\ wave\ (v) \times time\ (t)}{2}$$

Remember!
Sound travels faster through solids and liquids than gases, as the molecules are closer together.

G–E

D–C

Ultrasound treatment

Removing kidney and bladder stones

- Kidney and bladder stones are hard mineral deposits which can cause blockages and severe pain. They can be as large as 1 cm in diameter.

- Kidney and bladder stones can be destroyed using high frequency ultrasound.

- Pulses of ultrasound waves are focused onto the stones, and this high frequency vibration causes the stones to break up into small fragments which can easily pass out of the body naturally in the urine.

Physiotherapy

- High frequency ultrasound in human tissue causes fluctuations in pressure, which leads to heating in the body.

- In physiotherapy ultrasound can be used to produce localised heating in damaged muscle tissue, which increases the rate of healing.

Risks

- Very intense ultrasound can cause dissolved gases (oxygen and carbon dioxide) to form tiny bubbles in the blood.

- When the bubbles collapse they can damage tissue in the body.

- Medical uses of ultrasound always use low-intensity ultrasound.

B–A*

Improve your grade

Ultrasound

Higher: During an ultrasound scan the radiographer will use a jelly lubricant between the transducer and the skin. Using your knowledge of sound waves, explain why this is done.   *AO2* [3 marks]

Producing X-rays

Evacuated tubes

- An evacuated tube is a glass vessel with no air inside (a vacuum).

- If a potential difference is applied across an **anode** and a **cathode** in an evacuated tube, a beam of electrons can be forced to flow.

- A low voltage supply (6 V) heats up the filament at the cathode, and causes the electrons in the filament to gain enough energy to escape. This is known as **thermionic emission**.

- A high voltage supply (500 V) applied between the cathode and the anode causes the electrons to be accelerated towards the positively charged anode.

- As the electrons are moving, there is a flow of charge or electric current.

- If there is any air in the tube the electrons will collide with air molecules and lose energy, so it has to be a vacuum.

- This arrangement is sometimes known as a thermionic diode.

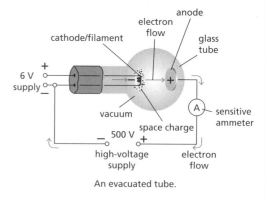

An evacuated tube.

X-ray tubes

- An X-ray tube is similar to a thermionic diode, but a higher potential difference (about 100 000 V) is needed between the anode and the cathode.

- The anode is made of a high melting point metal, such as tungsten, and is known as the **target**.

- When the electrons strike the target at very high speed, about 1% of their energy is converted to X-rays, with the rest converted to heat.

- As there is so much heat produced the target needs to be cooled by circulating water, otherwise the target may melt.

- The intensity of the X-rays produced can be increased by increasing the temperature of the filament (by increasing the current flowing in it).

- The frequency of the X-rays produced can be increased by increasing the accelerating potential difference between the cathode and the anode.

- The higher the frequency of the X-rays, the higher their energy and higher their ionising power.

An X-ray tube is really a thermionic diode.

> **Remember!**
> As the frequency of the radiation increases, its wavelength decreases.

Electron beams

Electric current

- In an electron beam:

  electric current (Amps) = number of particles per second (1/s) × charge on each particle (Coulomb, C)

  $$I = N \times q$$

- The charge on one electron is denoted e, and is equal to 1.6×10^{-19} C.

Kinetic energy

- The kinetic energy of the electron arriving at the target is equal to the electrical energy gained by accelerating through the potential difference.

  Kinetic energy (J) = charge on one electron (C) × accelerating potential difference (V)

  $$\frac{1}{2}mv^2 = e \times v$$

- The mass of one electron is 9.1×10^{-31} kg.

- The greater the kinetic energy of the electrons, the higher the frequency of the X-rays produced.

⦿ Improve your grade

Evacuated tubes
Foundation: Explain what is meant by the term 'thermionic emission'.

AO1 [3 marks]

Medical uses of X-rays

Using X-rays for diagnosis

CAT scans

- Computerised Axial Tomography (CAT) is used to diagnose:
  - Tumours
  - Blood clots
  - Alzheimer's disease.

- Unlike conventional 2-dimensional X-rays scans, CAT scans can also show the structure of soft tissue in the body. They produce a 3-dimensional image of the inside of the body.

- During a CAT scan X-rays are passed through the body from all directions as the X-ray source rotates round the body in a spiral path. Detectors produce images of slices of the body to build up a 3-dimensional image of the body.

- As X-rays are ionising radiation, they can damage healthy cells. Risks are minimised by using short exposure times.

Fluoroscopes

- Fluoroscopy is used to obtain real-time moving images of the internal structure of organs in the body, usually used to diagnose problems with the digestive system.

- The patient first has to drink a chalky fluid containing barium sulphate, which is opaque to X-rays.

- When X-rays are passed through the body to a detector, the passage of the 'barium meal' can be viewed on a monitor, and the doctors can identify any problems.

G–E

Absorption of X-rays

- The intensity of electromagnetic radiation decreases with distance from its source, and obeys an inverse square law with distance.
  - Intensity (I) is inversely proportional to the square of the distance (r) from the source.

$$I \propto \frac{1}{r^2}$$

- X-rays are absorbed by any material they pass through. For each different material the transmitted intensity decreases as the thickness of the material increases.

Remember!
Lead is a good absorber of X-rays, so radiographers wear lead-lined aprons to protect their bodies.

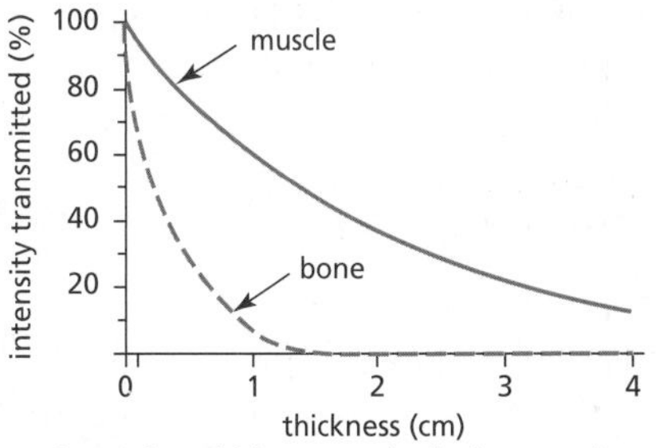

Intensity against thickness graphs for bone and muscle.

D–C

Using X-rays for treatment

- Radiotherapy is used for the treatment of cancer.

- Intense X-ray or gamma radiation is directed at a tumour to destroy the cancerous cells.

- A linear particle accelerator (linac) is used to produce X-rays, so that the energy and intensity of the X-rays can be easily controlled, to ensure that the patient receives the correct dose.

- X-rays produced from a linac can also be easily switched off, unlike gamma rays from radioactive sources.

B–A*

How science works

You should be able to:

- understand that new technologies may develop new methods for using X-rays in medicine
- analyse the risks and benefits of X-ray technology.

Improve your grade

Absorption of X-rays

Higher: Explain what is meant by an 'inverse square law' for the absorption of X-rays.　　　*AO1* [2 marks]

Heart action and ECG

ECG

- An electrocardiogram (ECG) is a recording of the electrical activity of your heart.

- Small metal electrodes are stuck onto arms, legs and chest with a conducting gel.

- The electrodes are then connected to a monitor which displays the electrical signals which make your heat beat.

- Normally a healthy heart beats regularly at about 70 beats per minute when resting.

- During exercise the heart rate will increase.

- Frequency (in Hertz) of the heartbeat is the number of cycles per unit time (second).

- The time period is the time taken for one complete cycle.

- $Frequency\ (Hz) = \dfrac{1}{Time\ period\ (s)}$

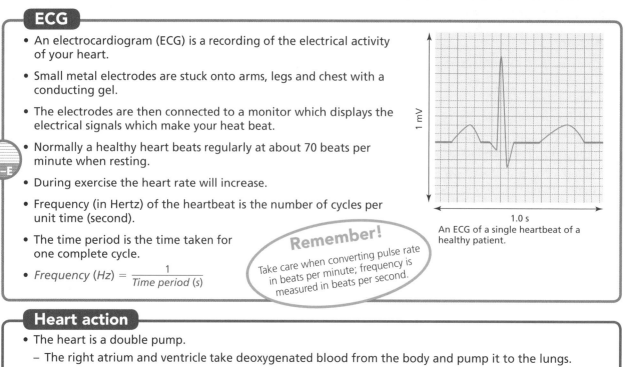

1.0 s

An ECG of a single heartbeat of a healthy patient.

Remember!
Take care when converting pulse rate in beats per minute; frequency is measured in beats per second.

G–E

Heart action

- The heart is a double pump.
  - The right atrium and ventricle take deoxygenated blood from the body and pump it to the lungs.
  - The left atrium and ventricle take the oxygenated blood from the lungs and pump it round the body.

- The regular pumping action is controlled by special muscle cells which are activated by changes in electrical potential known as action potentials of the heart.

- An ECG shows the changes in electrical potential and can be linked to the action of the heart:
  - P wave is when the atria contract
  - QRS spike is when the ventricles contract
  - T wave is when the ventricles relax.

Linking the action potential to the action of the heart. The activated muscles are shown in blue.

Structure of the heart.

- deoxygenated blood
- oxygenated blood

pulmonary artery (to the lungs) aorta (to the body)

vena cava (from the body)

pulmonary vein (from the lungs)

right atrium

left atrium

right ventricle

left ventricle

D–C

- Heart problems can be identified by studying the shape and frequency of an ECG.

- The frequency of the heartbeat depends on the demand from the body.

- The shape of the ECG trace is affected by the intake of stimulants such as caffeine.

- Doctors can prescribe drugs to regulate heart beats; for more severe problems a **pacemaker** is implanted.

B–A*

normal heartbeat

fast heartbeat (tachycardia)

slow heartbeat (bradycardia)

irregular heartbeat (arrhythmia)

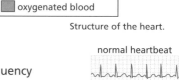

time (1 s)

You can identify heart (cardiac) problems by examining ECGs.

Improve your grade

Heart action and pacemakers

Higher: Explain when a doctor might implant a pacemaker in a patient's heart.

AO1 [2 marks]

Pulse oximetry

Pulse oximeters

- The average resting pulse rate for an adult is between 60 and 100 beats per minute. Elite athletes have a resting pulse rate of between 40 and 60 beats per minute.

- The maximum heart rate is 220 minus your age in years. The target for a healthy pulse rate during or just after exercise should be 60%–80% of this maximum.

- A pulse oximeter measures:
  - The pulse rate in beats per minute
  - The amount of oxygen in the blood.

- A pulse oximeter is small and is worn on the finger or earlobe.

- A pulse oximeter is regularly used:
  - In intensive care
  - During anaesthesia
  - To investigate sleep disorders.

Remember!
If the pulse rate is low, less oxygen will be pumped round the body.

G–E

- **Haemoglobin** in the blood is a protein which carries oxygen from the lungs to the organs of the body.

- The amount of light absorbed by haemoglobin varies depending on the amount of oxygen it carries.

- A pulse oximeter uses a bright light-emitting diode (LED) to shine light through the patient's finger. A light detector then records the amount of light transmitted through the finger.

- The difference in intensity between the emitted light and the transmitted light is used to determine the percentage of haemoglobin which is saturated with oxygen.

- Normally the blood should be 97% saturated with oxygen.

A finger in a pulse oximeter. The artery carries blood from the heart to the finger and the vein carries the blood back towards the heart.

D–C

- There are two different forms of haemoglobin molecule in the blood – oxyhaemoglobin (HbO_2) and reduced haemoglobin (Hb).

- The amount of light absorption of each type of haemoglobin depends on the wavelength of the light.

- Oxidised haemoglobin absorbs more infrared radiation than visible red light.

- Reduced haemoglobin absorbs more visible red light than infrared light.

- A pulse oximeter uses two LEDs, one emitting visible red light (wavelength 650 nm) and one emitting infrared radiation (wavelength 950 nm).

- The sensor of an oximeter is connected to a computer which can determine the proportion of each type of haemoglobin from the amount of each wavelength light which is absorbed.

B–A*

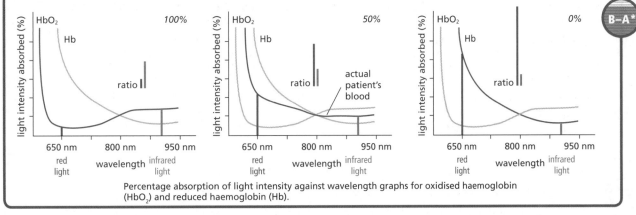

Percentage absorption of light intensity against wavelength graphs for oxidised haemoglobin (HbO_2) and reduced haemoglobin (Hb).

Improve your grade

Pulse rates

Foundation: Why is it important to monitor a patient using pulse oximetry when he or she is under anaesthetic?   *AO2* [2 marks]

Ionising radiations

The nuclear atom

- All materials are made up of atoms.

- In the centre of an atom is a tiny positively charged nucleus, which contains nucleons (neutrons and protons).

- Surrounding the nucleus are negatively charged electrons.

- The nucleus of an atom is represented as $^A_Z X$:
  - X is the chemical symbol for that element,
  - Z is the total number of protons inside the nucleus and is called the **atomic** or **proton number**,
  - A is the total number of neutrons and protons in the nucleus and is called the **mass** or **nucleon number**.

- The **isotopes** of an element are nuclei which have the same number of protons, but differing numbers of neutrons.

- Atoms are neutral, as they have an equal number of electrons and protons, but ions are atoms which have either lost or gained electrons, so are electrically charged.

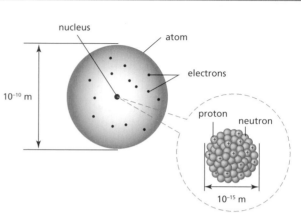

Model of the atom.

A summary of the masses and charges of the particles of the atom.

	Neutron	Proton	Electron
Charge relative to the proton	0	1	−1
Mass relative to the proton	1	1	0.00055

Nuclear radiations

- Some nuclei are unstable, and emit particles or electromagnetic radiations, all of which tend to be ionising radiations.

- The three main types of nuclear radiation are alpha (α) particles, beta (β) particles and gamma (γ) rays – these are emitted by radioactive isotopes.

- Some heavier isotopes can emit neutrons.

- Some lighter isotopes can emit positrons. A positron has the same mass as an electron, but has a positive charge of 1.6×10^{-19} C. It is the antiparticle to an electron, and is known as beta-plus (β^+).

> **Remember!**
> The charge on one electron is 1.6×10^{-19} C

Type of radiation	Nature of radiation	Charge	Properties
Alpha (α)	Helium nucleus ^4_2He	$+2e$	Strongly ionising, stopped by paper, skin or 6 cm air.
Beta (β^-)	Electron $^0_{-1}e$	$-e$	Weakly ionising, stopped by a few mm of aluminium.
Gamma (γ)	Short wavelength electromagnetic radiation	No charge	Very weakly ionising, significantly absorbed by a few cm of lead, or a few m of concrete.
Neutron	^1_0n	No charge	Does not interact much with matter.
Beta-plus (β^+)	Positron ^0_1e	$+e$	A positron interacts strongly with an electron to produce gamma rays.

- Radioactive materials are used in the treatment and diagnosis of cancer.

- Ionising radiations can damage living cells, so exposure is kept to a minimum.

- Radiographers and other people who work with ionising radiation need to wear protective clothing, usually made of lead. They also wear dosimeters to record the amount of radiation exposure.

Improve your grade

The nuclear atom

Foundation: Carbon-14 is a radioactive isotope which is written $^{14}_6$C. How many protons and neutrons are present in the nucleus?

AO1 [2 marks]

Radioactive decays

Alpha (α) decay

- An alpha particle contains 2 protons and 2 neutrons.
- An unstable nucleus of heavier atoms may emit alpha particles, and transmute into a new daughter nucleus.
- The daughter nucleus will have a proton number which is reduced by 2 and a mass number reduced by 4.
- For example: Uranium-238 ($^{238}_{92}U$) is an alpha emitter, and has 92 protons and 238 nucleons.
  - Its daughter nucleus will have 90 protons and 234 nucleons.
  - Using the periodic table, an element with an atomic number of 90 is thorium so the new nucleus will be thorium-234 ($^{234}_{90}Th$).

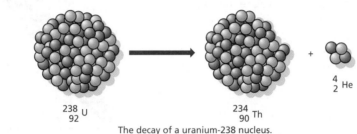

$$^{238}_{92}U \qquad ^{234}_{90}Th \qquad ^{4}_{2}He$$

The decay of a uranium-238 nucleus.

G–E

Beta-minus (β⁻) decay

- Some lighter unstable nuclei emit beta particles, which are fast moving electrons.
- A neutron (inside the nucleus) changes into a proton and an electron; the electron is emitted as a beta particle, and the proton remains in the nucleus.
- The atomic number of the daughter nucleus will therefore increase by +1, and the nucleon number remains the same.
- For example: Carbon-14 is a beta emitter, and has 6 protons and 14 nucleons.
  - The daughter nucleus will have 7 protons and 14 nucleons.
  - Using the periodic table, an atom with an atomic number of 7 is nitrogen, so the new nucleus is nitrogen-14 ($^{14}_{7}N$).

Remember!
During α, β⁻ and β⁺ decays the nucleus will always change into a different element, but γ emission will not.

D–C

Gamma (γ) decay

- Gamma rays are often emitted following alpha or beta emission, to reduce surplus energy.
- Gamma rays have no charge, and so there is no change to the structure of the nucleus. The nucleus undergoes some rearrangement to get to a lower energy state.

D–C

Beta-plus (β⁺) decay

- Some proton-rich nuclei undergo positron (beta-plus) decay.
- A positron removes a positive charge from the nucleus. A proton changes into a neutron and a positron.
- The atomic number of the daughter nucleus will decrease by 1, and the mass number remains the same.
- This can be shown in a nuclear reaction for the beta-plus decay of oxygen-15:

$$^{15}_{8}O \longrightarrow {}^{15}_{7}N + {}^{0}_{+1}e.$$

- All radioactive decays can be shown as nuclear reaction equations:
  - the sum of the mass number on each side of the equation must be equal, and
  - the sum of the atomic number on each side of the equation must be equal.
- Alpha decay of uranium-238: $^{238}_{92}U \longrightarrow {}^{234}_{90}Th + {}^{4}_{2}He.$
- Beta minus decay of carbon-14: $^{14}_{6}C \longrightarrow {}^{14}_{7}N + {}^{0}_{-1}e.$

B–A*

Improve your grade

Beta-minus decay

Higher: A beta-minus particle is a fast moving electron. Explain how an electron can be emitted from the nucleus, which only contains protons and neutrons. *AO1* [2 marks]

Stability of nuclei

N–Z curve

B–A*

- The N–Z curve is a graph of the number of neutrons against the number of protons for all known nuclei.
- The stable nuclei lie on a curve starting from the origin with slightly increasing gradient (the black line on the graph).
- Stable nuclei with atomic numbers less than about 20 have equal numbers of neutrons and protons.
- Most stable nuclei have more neutrons than protons.
- Unstable nuclei above the stability curve are 'neutron-rich' and those below the line are 'neutron-poor'.
- The protons and neutrons in the nucleus are held together by the **strong nuclear force**, which has to overcome the repulsion between the positively charged protons.

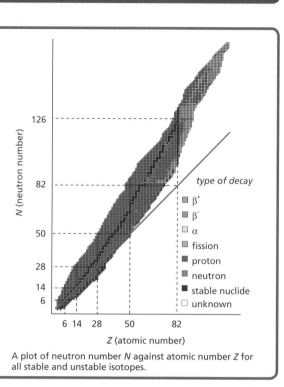

A plot of neutron number N against atomic number Z for all stable and unstable isotopes.

Beta-minus, beta-plus or alpha decay?

B–A*

- A neutron-rich nucleus (above the line) has too many neutrons to be stable, and decays by emitting a beta-minus (electron).
  - A neutron changes into a proton and an electron.
  - The atomic number of the daughter nucleus increases by one and the neutron number decreases by one.
  - This brings the nucleus closer to the stability curve.

$$^{23}_{10}Ne \longrightarrow ^{23}_{11}Na + ^{0}_{-1}e.$$

- A neutron-poor nucleus (below the line) has too many protons to be stable, and decays by emitting a beta-plus (positron).
  - A proton changes into a neutron and a positron.
  - The atomic number of the daughter nucleus decreases by one, and the neutron number increases by one.
  - This brings the nucleus closer to the stability curve.

$$^{19}_{10}Ne \longrightarrow ^{19}_{9}F + ^{0}_{+1}e.$$

> ### EXAM TIP
> Always check that both the atomic numbers and the nucleon numbers balance on both sides of all nuclear reaction equations.

- Unstable nuclei with a high atomic number (Z > 82) decay by emitting alpha particles.
  - As an alpha particle consists of 2 protons and 2 neutrons this has little effect on the nucleus' position relative to the stability curve.

Neutrinos and beta decays

B–A*

- During beta-plus decay, a nucleus also emits a neutrino.
- Neutrinos have no charge and have negligible mass (one billionth the mass of a proton), so they do not interact with matter, and are therefore very difficult to detect.

$$^{15}_{8}O \longrightarrow ^{15}_{7}N + ^{0}_{+1}e + \nu.$$

  - The Greek letter ν (nu) is used to denote a neutrino.
- The existence of neutrinos was predicted by Wolfgang Pauli in 1930, but they were only discovered 26 years later.
- The universe is believed to be saturated with neutrinos.

⦿ Improve your grade

Type of decay

Higher: Silicon-27 is an unstable isotope of silicon. Silicon has an atomic number of 14. Use your knowledge about stability of nuclei to work out which type of beta decay it will undergo. Explain your answer.
AO3 [3 marks]

Quarks

Quarks

- Quarks are tiny particles which make up all matter.
- The quark model was proposed independently by both Murray Gell-Mann and George Zweig in 1964.
- Quarks are fundamental particles – they cannot be subdivided into smaller particles.
- Electrons, positrons and quarks are fundamental particles.
- Protons and neutrons are made up of two types of quark – the up quark and the down quark.
  - The up quark has a charge of $+^2/_3 e$.
  - The down quark has a charge of $-^1/_3 e$.
- Experiment and theory both show that it is impossible to isolate quarks.
- Neutrons and protons are both made up of three quarks.
  - A neutron is made up of one up (u) quark and two down (d) quarks *Neutron = udd*
  - A proton is made up of two up quarks and one down quark. *Proton = uud*
- The total mass of a proton or neutron should be equal to the sum of the masses of the quarks they contain; however, the mass of the proton or neutron is about 80 times the mass of the quarks. This is explained by using Einstein's equation $E = mc^2$ and the equivalence of mass and energy.

proton   neutron

The arrangement of quarks inside a proton and a neutron.

B–A*

Quarks and beta decay

- Beta-minus decay can be explained in terms of quarks.
  - A down quark in a neutron changes into an up quark and an electron.
  - The electron is emitted as a beta-minus particle (so-called to emphasise that it is emitted from the nucleus and to distinguish it from an orbital electron), so the 'neutron' now has two up quarks and one down quark, so is now a proton.
- Beta-plus decay can also be explained in terms of quarks.
  - An up quark in a proton changes into a down quark and a positron.
  - The positron is emitted as a beta-plus particle (so-called to emphasise that it is emitted from the nucleus), so the 'proton' now has two down quarks and one up quark, so is now a neutron.
- A proton on its own is totally stable. A proton in the nucleus may change through beta-plus decay into a neutron and a positron.
- A neutron on its own can decay to release an electron. The half-life of a free neutron is about 11 minutes.

beta-minus decay
(neutron → proton + electron)

beta-plus decay
(proton → neutron + positron)

B–A*

How science works

You should understand about how theories and models are developed to explain scientific phenomena.

- In the 1930s scientists explained the existence of all matter in terms of three particles: electrons, protons and neutrons.
- In the 1950s particle accelerators produced many new particles which could not be explained with this model of matter.
- In the 1960s these new particles were explained using the quark model.
- Nowadays, scientists believe that quarks are the building blocks for all matter.

Improve your grade

Fundamental particles

Higher: Explain why electrons are said to be fundamental particles, but protons are not.   AO1 [2 marks]

Dangers of ionising radiations

Dangers of radiation

G–E

- Ionising radiation can damage body tissue. Cells may mutate, die or fail to reproduce themselves.
- Effects of radiation damage include skin burns, nausea, destruction of bone marrow, hair loss, sterility, cancers, changes to genetic material.
- People who work with ionising radiation are protected by:
  - Keeping the distance between the workers and the source of radiation as large as possible
  - Keeping the exposure time to a minimum
  - Using shielding, such as lead-lined aprons or concrete walls
  - Wearing dosimeters, which consist of radiation-sensitive film within a holder. The film goes darker when exposed to increasing amounts of radiation.

> **Remember!**
> We are being exposed to ionising radiation all the time from background radiation.

D–C

- There are three types of damage caused to cells by ionising radiation:
  - Cells become damaged, then repair themselves and operate normally
  - Cells become damaged, then repair themselves but function abnormally, such as failing to reproduce themselves, or reproducing at an uncontrolled rate; this is the main cause of cancers
  - Cells become so damaged they die.
- The amount of damage caused by the radiation depends on:
  - The dose received
  - The parts of the body exposed – rapidly dividing cells (such as blood cells and hair follicles) are most susceptible
  - The nature of the radiation – alpha particles are easily stopped by the skin, so do not present a serious problem unless the source gets inside the body.

Equivalent dose

B–A*

- The effective biological damage to human tissue by ionising radiation is known as the 'equivalent dose' and is measured in Sieverts (Sv).
- The higher the equivalent dose, the greater the chance of biological damage.
  - Radiation dose up to 0.25 Sv a day causes no acute symptoms
  - Radiation dose between 0.25 Sv and 1 Sv a day can cause nausea, loss of appetite and damage to bone marrow.
- The maximum permissible dose for medical personnel is 20 mSv per year, averaged over 5 years, with a maximum of 50 mSv in any one year.
- Patients are always given the smallest dose of radiation needed for their treatment.

Source of radiation	Equivalent dose (mSv)
Dental X-ray	0.0005
Full body CAT scan	15
Fluoroscopy	5
Estimated maximum dose to residents of Fukushima following the nuclear accident in 2011	68
Average background radiation	2

How science works

You should understand how ideas about the uses of radioactivity have changed over time, as we have now become more aware of the risks. For example:

- radium pendants were once worn at all times to cure rheumatism;
- radioactive water was drunk for good health and radioactive thorium was used in face creams.

Improve your grade

Damage to cells

Higher: Explain why blood cells and hair follicles are more susceptible to damage by ionising radiation.

AO2 [2 marks]

Treatment of tumours

Palliative radiotherapy

- Palliative radiotherapy is treatment to shrink a cancer or slow down its growth.
- Palliative radiotherapy does not aim to cure the cancer completely.
- Palliative radiotherapy uses either external or internal radiation.
- Palliative radiotherapy is painless, but it does have some side effects:
  - Tiredness
  - Sickness following radiation to the stomach, abdomen or brain
  - Feeling very sore following radiation to the lung, head or neck.
- Patients are often prescribed anti-emetic drugs to help control sickness.
- Not all cancers respond well to radiotherapy.

Remember!
Radiotherapy causes damage to cells because the radiation ionises the atoms in the cells.

G–E

External radiotherapy

- External radiotherapy works well for treating cancer cells in a localised area of the body.
- Usually X-rays from a linear accelerator (linac) are used (rather than gamma radiation from a radioactive source), so that the energy and intensity of the radiation are easily controlled, and aimed at the specific area of the body.
- Patients can have external radiotherapy for more than one area of the body at the same time.
- Patients can have 1 or 2 treatments or up to 10 short treatments given over 2 weeks.

D–C

Internal radiotherapy

- There are several different types of internal radiotherapy:
  - Patients may have a small injection of a radioactive substance to treat widespread cancer in the bones
  - Patients may have a radioactive metal implant put inside the body very close to the cancer. These are sometimes in the form of wires
  - Patients may be given a drink or capsule containing radioactive materials.
  - After being treated with radioactive materials, patients are often allowed to go home, but are advised not to go near children or pregnant women.

D–C

Using neutrons

- An intense beam of fast moving neutrons can also be used in the treatment of cancer.
- Neutrons have no charge, but because they have a large speed and mass they have enormous amounts of kinetic energy and can ionise atoms.
- Neutrons can produce 5 times more ionisation than X-rays and beta particles.
- Intense neutron beams are produced by bombarding beryllium with protons from a particle accelerator.
- The neutron beam is aimed at the tumour with the help of lasers, and is able to stop the growth of the tumour, or possibly destroy it completely.

B–A*

How science works

You should understand that there are social and ethical issues associated with cancer treatments.

- Cancer treatments are very painful; yet there is currently no cure for cancer.
- Cancer treatments are very expensive; so should they be used to ease suffering despite the fact there is no cure?

Improve your grade

Palliative radiotherapy

Foundation: Palliative radiotherapy is often given to cancer sufferers. Describe one benefit and one drawback of this type of therapy. *AO1* [2 marks]

Diagnosis using radioactive substances

Radioactive tracers

- A medical radioactive tracer is a radioactive substance that is either injected into or swallowed by a patient.
- Tracers are absorbed differently by different tissues in the body; a radiation detector can produce an image showing the distribution of the radioactive substance.
- Radioactive tracers are either beta or gamma emitters with short half-lives.

Tracer	Symbol	Radiation	Half-life	Used to monitor
Iodine-131	$^{131}_{53}I$	Beta	8.1 days	Passage of fluid through the kidney
Technetium-99	$^{99}_{43}Tc$	Gamma	6 hours	Blood flow in the brain or lungs; growth of bones; blood circulation in the heart
Xenon-133	$^{133}_{54}Xe$	Gamma	2.3 days	Function of the lungs

- For example Iodine-131 is injected into a patient with kidney problems:
  - A healthy kidney will pass the iodine through to the bladder
  - If there is a blockage, the iodine builds up in the kidney and can be detected.

Remember!
The radioactive tracer needs to have a suitable half-life – not so short that there is no time to see the effect, but not too long either, so it does not exist for too long in the body.

Positron emission tomography (PET) scanners

- PET scanners are used to monitor:
  - Activity of the brain
  - Spread of cancer through the body
  - Flow of blood through organs such as the heart.
- Radioactive isotopes (beta-plus emitters) are first produced in a **cyclotron** (particle accelerator).
- The radioactive isotopes are attached (or 'tagged') to natural chemicals found in the body, such as glucose, water and ammonia. This tagged substance is known as a **radiopharmaceutical** and is injected into the patient.
- When the radiopharmaceutical is injected into the body it will go to areas which use the natural chemical. For example:
  - Fluorine-18 is tagged to glucose to produce fluorodeoxyglucose (FDG)
  - Cancers use glucose differently from normal tissues so the FDG will reveal cancerous tissues.
- The radioactive isotopes used include carbon-11, nitrogen-13, oxygen-15 and fluorine-18, which have short half-lives, so have to be produced just before use. This means that the cyclotron has to be located within the hospital grounds.
- A PET scanner detects the gamma rays produced by the annihilation of positrons (emitted by the radioactive isotopes) and electrons. *More about how PET scanners work on page 63.*

- PET scanners are very expensive because they also require a cyclotron to produce the radiopharmaceuticals.
- There are only about 150 PET scanners around the world, mostly in the developed world – USA, Europe and Japan.

Improve your grade

Radioactive tracers

Foundation: There are many radioactive isotopes of iodine. Iodine-131 is a beta emitter with a half-life of about 8 days. Iodine-108 is an alpha emitter with a half-life of 36 ms. Explain why Iodine-108 cannot be used as a radioactive tracer. *AO3* [3 marks]

Particle accelerators and collaboration

Particle accelerators

- Particle physicists believe all matter to be made up of two fundamental groups of particles – quarks and leptons.
  - Neutrons and protons are made up of quarks.
  - Electrons and neutrinos are examples of leptons.
- Physicists study the collisions between high-speed electrons, protons and ions with matter to investigate the structure and behaviour of particles.
- Particle accelerators use electric and/or magnetic fields to accelerate charged particles almost up to the speed of light.
  - Linear accelerators are long and straight. The longest one is over 3 km long.
  - Some accelerators are circular, so that the charged particles accelerate around several times to increase their velocity.
  - Small circular accelerators (called cyclotrons) are used in hospitals to make radioactive isotopes for PET scanners.

> **Remember!**
> In order to work out the speed of a particle travelling in a circle you need the circumference, which is equal to $2\pi r$

G–E

- The European Organisation for Nuclear Research (CERN) is an international organisation whose main purpose is to provide particle accelerators for high energy particle physics research.
- CERN's facilities include the world's largest accelerator – the Large Hadron Collider (LHC), and are located on the French–Swiss border.
- The LHC:
  - Is the most expensive particle accelerator in the world, with a budget of about €8 billion
  - Collides protons and ions at higher speeds and energies than has ever been possible before
  - Accelerates particles in a circular path of circumference 27 km and up to 11000 revolutions per second
  - Uses superconducting electromagnets at very low temperatures to produce very strong magnetic fields, which are used to accelerate the particles.
- It is hoped that the work completed with the LHC will help to answer some of the fundamental questions about the structure of matter, about events just after the Big Bang and also explain why particles have mass.

D–C

Collaboration

- CERN is a pure research organisation, and the work it completes may never have any commercial value.
- CERN's constitution states that it must do no work for military purposes, and all the results of all its experimental and theoretical work must be published and made available to the general public.
- CERN is an example of collaborative international research.
- Collaborative international research has many advantages, including:
  - Sharing the cost between many countries and organisations, as the running costs are too high for any one organisation
  - There are no political barriers to communicating new understanding about the physical world
  - Speeding up progress by sharing ideas among many scientists.

B–A*

How science works

You should understand that there are some questions about the structure of matter which cannot currently be answered.

- Research institutions like CERN are hoping to answer some of these questions.

Improve your grade

Particle accelerators

Foundation: Particle accelerators can either be linear or circular. Describe one advantage of circular particle accelerators over linear ones. *AO2* [2 marks]

Cyclotrons

Circular motion

- An object moving in a circle at constant speed is continuously changing its direction of motion.
  - Its velocity is constantly changing, therefore it is accelerating.
  - The direction of the acceleration and the resultant force providing that acceleration is towards the centre of the circle. The force is known as the **centripetal force**.
  - The centripetal force is always at right angles to the velocity and is always directed towards the centre of the circle.

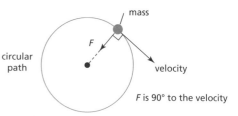

For an object to move in a circle there must be a centripetal force *F* acting on the object towards the centre of the circle.

- A charged particle moving within a magnetic field will experience a force which is always at right angles to its velocity. This will be a centripetal force.
  - The centripetal force will make the charged particle go around in a circular path.
  - In a given magnetic field the radius of the circle will increase as the speed of the particle increases.
  - Negative and positive charges will move in opposite directions in a magnetic field.

Remember!
Velocity is a vector quantity – it has both direction and size. Particles move in a circle at constant speed but are changing direction all the time, so the velocity all is changing.

G–E

Cyclotrons

- A cyclotron uses a magnetic field to accelerate charged particles in a circular path.
  - It has two hollow D-shaped metal cavities called 'dees', which are enclosed in an evacuated chamber.
  - Powerful electromagnets are used to provide a uniform magnetic field at right angles to the plane of the dees.
  - A source of particles (electrons, protons etc) is placed at the centre of the cyclotron.
  - The dees are connected to a very high frequency alternating voltage supply which repeatedly reverses the polarity of the dees.

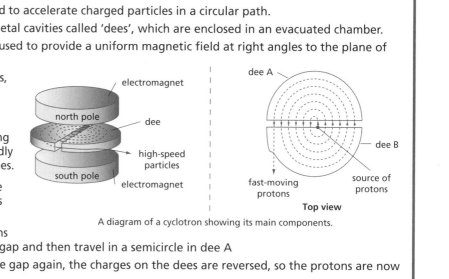

A diagram of a cyclotron showing its main components.

- For example, if the source at the centre of the cyclotron produces positively charged protons:
  - If dee A is negative the protons will be accelerated across the gap and then travel in a semicircle in dee A
  - When the protons arrive at the gap again, the charges on the dees are reversed, so the protons are now accelerated towards dee B
  - Their speed has thus increased so they now travel in a larger diameter circular path in dee B
  - This process is repeated many times before the proton exits at very high speed.
- Cyclotrons are used to produce radioactive isotopes for medical purposes such as PET scanners.
- Stable isotopes can be bombarded with protons which have been accelerated in cyclotrons.

D–C

- For example if oxygen-18 (in enriched water) is bombarded with protons, fluorine-18 is produced.
  - $^{1}_{1}p + {}^{18}_{8}O \longrightarrow {}^{18}_{9}F + {}^{1}_{0}n + \gamma$
  - The neutron produced in the reaction is not harmful, but dangerous high-energy gamma rays are also produced in this bombardment.
- Many of the isotopes produced in cyclotrons have short half-lives, so the cyclotron needs to be close to the site of the PET scanner.

B–A*

Improve your grade

Cyclotrons

Higher: In a cyclotron similar to the one in the diagram above, the particles describe a spiral path. Explain how the magnitude and direction of the particle's velocity changes during this spiral path.

AO2 [3 marks]

Electron–positron annihilation

Matter, antimatter and annihilation

- Ordinary matter consists of neutrons, protons and electrons.

- Particles of **antimatter** have the opposite charge of the particles of matter.
  - A positron has identical mass to an electron, but has equal and opposite charge.
  - An anti-proton has the same mass as a proton, but equal and opposite charge.

- When a positron and an electron collide they destroy each other completely. This is known as annihilation.

- The total mass of both particles is converted into energy in the form of gamma rays. **Mass-energy** is conserved.

- There are two gamma rays which fly off in opposite directions so that **momentum** is also conserved.

- Gamma rays are electromagnetic radiation and have no charge. The total charge during the annihilation also remains constant.

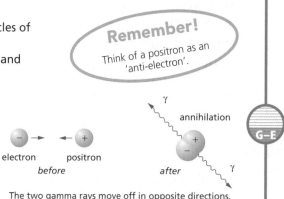

Remember!

Think of a positron as an 'anti-electron'.

electron   positron
before

annihilation

after

The two gamma rays move off in opposite directions.

G–E

PET scanners

Remind yourself about PET scanners on page 60.

- The patient is placed in a ring of gamma ray detectors which are connected to computers.

- The positrons emitted by the tracers collide with electrons to produce two gamma ray pulses in opposite directions.

- The difference in the arrival times and location of the detection of the two gamma rays is analysed by the computer to work out the exact location of the annihilation, and hence the radioactive tracer can be pinpointed.

- The computer builds up a 3-dimensional image of the distribution of the tracer in the body.

- PET scanners are used to study cancer, Alzheimer's disease, dyslexia and epilepsy.

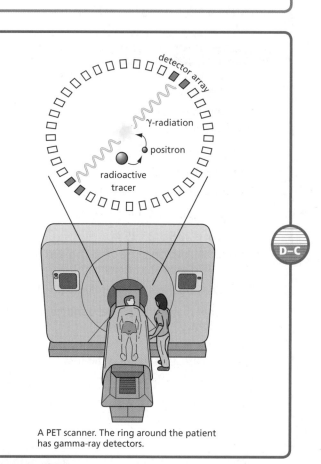

detector array

γ-radiation

positron

radioactive tracer

A PET scanner. The ring around the patient has gamma-ray detectors.

D–C

Energy released in electron–positron annihilation

- During electron–positron annihilation the total mass of the two particles is converted to energy.

- The energy released (E) is calculated using Einstein's **mass-energy equation**.

  $E = m c^2$

  m (total mass of the particles) $= 9.11 \times 10^{-31} + 9.11 \times 10^{-31}$ kg

  c (the speed of light) $= 3.0 \times 10^8$ m/s.

  $E = (1.82 \times 10^{-30}) \times (3.0 \times 10^8)^2 = 1.64 \times 10^{-13}$ J.

B–A*

🔘 Improve your grade

Matter, antimatter and annihilation

Foundation: During a collision between an electron and a positron, explain what physical quantities are conserved.

AO1 [3 marks]

Momentum

Conservation of momentum

- Momentum is a vector quantity as it has both direction and magnitude.
- *Momentum (kg m/s) = mass (kg) × velocity (m/s)*.
- The **principle of conservation of momentum** can be applied to all collisions (when there are no external forces acting on the objects).
  - The total momentum of the objects before the collision = the total momentum of the objects after the collision.
- In the diagram:

  Total momentum before collision
  = (10 kg × 4.0 m/s) + (5 kg × −2.0 m/s)
  = 40 kg m/s − 10 kg m/s = 30 kg m/s.

  Total momentum after collision
  = (10 kg × 1.5 m/s) + (5 kg × 3 m/s)
  = 15 kg m/s + 15 kg m/s = 30 kg m/s.

G–E

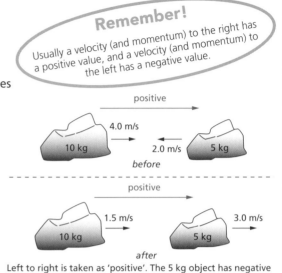

Remember!
Usually a velocity (and momentum) to the right has a positive value, and a velocity (and momentum) to the left has a negative value.

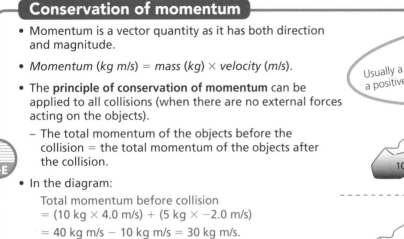

Left to right is taken as 'positive'. The 5 kg object has negative velocity and hence negative momentum.

Elastic and inelastic collisions

D–C

- There are two types of collision – **elastic collisions** and **inelastic collisions**.
- In the collision shown in the diagram above:

  Kinetic energy before the collision
  $$= \left(\frac{1}{2} \times 10.0 \times 4.0^2\right) + \left(\frac{1}{2} \times 5.0 \times 2.0^2\right)$$
  = 80 + 10 = 90 J

  Kinetic energy after the collision
  $$= \left(\frac{1}{2} \times 10.0 \times 1.5^2\right) + \left(\frac{1}{2} \times 5.0 \times 3.0^2\right)$$
  = 11.25 + 22.5 = 33.75 J

  - The total kinetic energy of the objects has not been conserved (56.25 J has been transformed into other forms such as heat and sound).

- When a ball bounces on the ground it undergoes an inelastic collision, as it will not regain its original height.
- Factors that may affect the rebound height include the material of the ball, the type of surface and the initial height.

	Elastic collision	Inelastic collision
Total momentum is conserved	✔	✔
Total energy is conserved	✔	✔
Kinetic energy is conserved	✔	✘

Elastic collisions and inelastic collisions compared.

B–A*

- In a totally inelastic collision the two objects will stick together and move off as one object with a common velocity *v*.

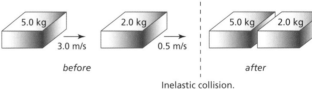

Inelastic collision.

- The velocity *v* can be calculated using the principle of conservation of momentum.

  Total momentum before collision = total momentum after collision
  (5.0 × 3.0) + (2.0 × 0.5) = (5.0 + 2.0) × *v*
  15.0 + 1.0 = 7 *v*
  v = 16/7 = 2.29 *m/s*

⊙ How science works

You should be able to investigate the factors affecting the rebound height of a ball.

⊙ Improve your grade

Inelastic collisions

Higher: An object of mass 150 g travels in a straight line with velocity 4 m/s and collides with another object with mass 100 g which is moving in the opposite direction with a velocity 3 m/s. The two objects stick together. Calculate their common velocity. *AO2* [3 marks]

Matter and temperature

Kinetic theory of matter

- The kinetic theory of matter:
  - Describes the behaviour of the three states of matter (solid, liquid and gas) in terms of the movement of particles
  - Explains how the average kinetic energy of the particles is related to temperature
  - Explains why gases exert pressure.

increasing temperature →

	Solid	Liquid	Gas
Arrangement of particles	Regular pattern Particles are very close together	Random arrangement Particles are close together	Random arrangement On average the particles are far apart
Movement of particles	Particles vibrate about fixed positions. As temperature increases, they vibrate more quickly	Particles move around each other. As temperature increases, their average speed increases	Particles move quickly in all directions and have a variety of speeds (**random motion**)

The three states of matter.

G–E

Temperature scales

- At absolute zero (0 K) all particles of substance stop moving and have zero kinetic energy.

- The Kelvin scale of temperature starts at absolute zero, and a temperature change of 1 °C is the same as a temperature change of 1 K.

- Absolute zero is the same as −273 °C.
  - To convert temperature in °C to temperature in K, you add 273
  - To convert temperature in K to temperature in °C you subtract 273
  - For example 25 °C will be (25 + 273) = 298 K.

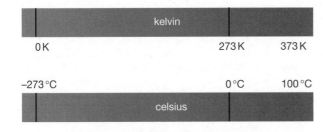

A comparison of the Kelvin and celsius scales.

D–C

Heat and temperature

- When heat energy is supplied to a gas, the particles move faster, and therefore collide more frequently with each other and the sides of their container.

- The average kinetic energy per particle increases, and this increases the total kinetic energy of the gas.

- The temperature (T) of a gas in Kelvin is proportional to the average kinetic energy (E) of the particles.
  - $E \propto T$

B–A*

Improve your grade

Kinetic theory of matter

Foundation: Describe the arrangement and movement of particles in a liquid as its temperature increases.

AO1 [3 marks]

Investigating gases

Pressure

- Atmospheric pressure is caused by air molecules colliding with every surface.
- Each molecule makes repeated collision with the wall X of the container.
- Each molecule exerts a tiny force on this wall. 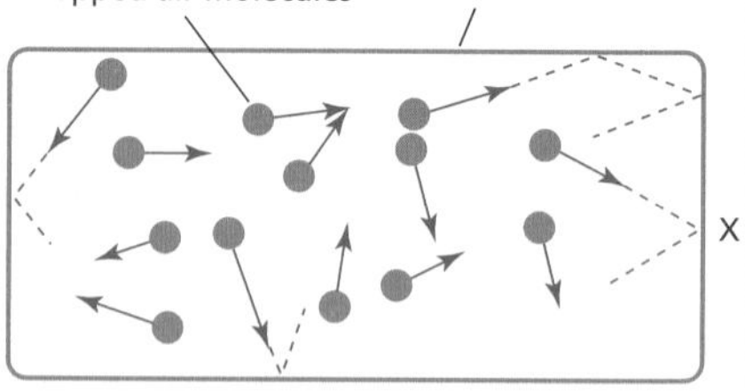
- $\text{Pressure on wall X (Pa)} = \dfrac{\text{total force exerted by all the molecules (N)}}{\text{area of wall X (m}^2\text{)}}$.

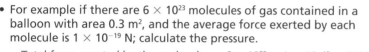

trapped air molecules   sealed container

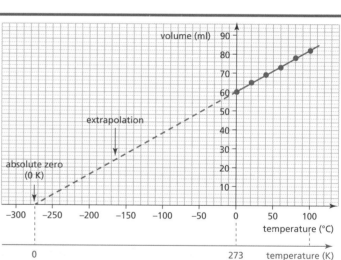

Pressure is caused by repeated collisions of molecules with the container walls.

- For example if there are 6×10^{23} molecules of gas contained in a balloon with area 0.3 m², and the average force exerted by each molecule is 1×10^{-19} N; calculate the pressure.

    Total force exerted by the molecules = $6 \times 10^{23} \times 1 \times 10^{-19}$ = 60000 N

    Pressure = 60000/0.3 = 20000 Pa

- The unit of pressure is the Pascal and has the unit Pa. It can also be written as Newton per square metre (N/m²).
- Atmospheric pressure is about 100 000 Pa (10^5 Pa).

G–E

Boyle's Law

- The pressure P exerted by a fixed mass of gas at constant temperature is inversely proportional to its volume V.
- $P \propto \dfrac{1}{V}$ or $P\,V = constant$

    $P_1\,V_1 = P_2\,V_2$

    – where V_1 and P_1 are the initial volume and pressure of the gas, and V_2 and P_2 are the final volume and pressure.

D–C

EXAM TIP

When quantities are inversely proportional to each other, if you double one quantity the other quantity will halve.

Charles' Law

- The volume V occupied by a fixed mass of gas kept at constant pressure is directly proportional to its temperature.
- $V \propto T$ or $\dfrac{V}{T} = constant$

    $\dfrac{V_1}{T_1} = \dfrac{V_2}{T_2}$

- Charles' Law can be investigated using a syringe, with its nozzle sealed. The syringe full of gas can be immersed in some boiling water.
- As the water cools the temperature and volume of the gas can be recorded and used to plot a graph.
- The graph shows a straight line, and can be extrapolated back to show that at absolute zero (−273 °C) the volume of the gas is theoretically zero.

A graph of volume of a fixed mass of gas against temperature in °C. You can determine absolute zero from this graph.

B–A*

- For example 0.5 m³ of gas at a constant pressure, cools down from 80 °C to room temperature (25 °C). Calculate the final volume it occupies.
    - First of all write down all the figures you have been given:

        V_1 = 0.5 m³ T_1 = 273 + 80 = 353 K

        V_2 = ? T_2 = 273 + 25 = 298 K

    - Substitute the figures into the equation to solve for V_2

        $\dfrac{0.5}{353} = \dfrac{V_2}{298}$

        $0.00142 = \dfrac{V_2}{298}$

        $V_2 = 0.00142 \times 298 = 0.42$ m³

Boyle's Law

Higher: When you decrease the volume of a fixed mass of gas at a constant temperature, its pressure increases. Explain this using ideas about the kinetic theory of gases.   *AO2* [3 marks]

The gas equation

The pressure law

- As the temperature of a gas increases, the molecules' speed increases.
  - So the molecules collide with (each other and) the walls more frequently
  - This causes the force and therefore the pressure on the walls to increase.
- The **pressure law** states that the pressure P, exerted by a fixed mass of gas, kept at constant volume, is directly proportional to its temperature T on the Kelvin scale.

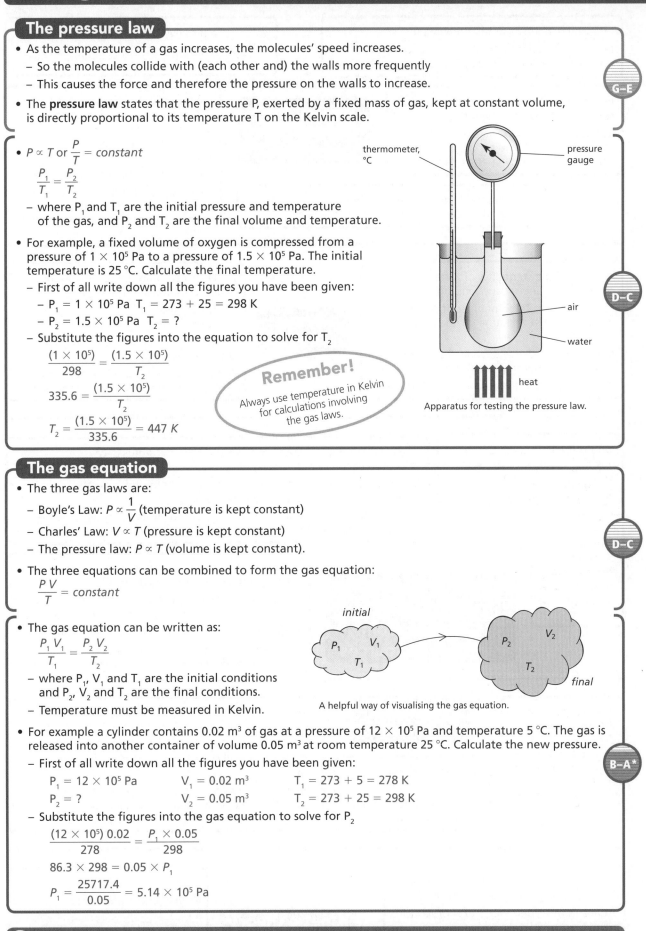

- $P \propto T$ or $\dfrac{P}{T} = constant$

  $\dfrac{P_1}{T_1} = \dfrac{P_2}{T_2}$

  - where P_1 and T_1 are the initial pressure and temperature of the gas, and P_2 and T_2 are the final volume and temperature.

- For example, a fixed volume of oxygen is compressed from a pressure of 1×10^5 Pa to a pressure of 1.5×10^5 Pa. The initial temperature is 25 °C. Calculate the final temperature.
  - First of all write down all the figures you have been given:
    - $P_1 = 1 \times 10^5$ Pa $T_1 = 273 + 25 = 298$ K
    - $P_2 = 1.5 \times 10^5$ Pa $T_2 = ?$
  - Substitute the figures into the equation to solve for T_2

    $\dfrac{(1 \times 10^5)}{298} = \dfrac{(1.5 \times 10^5)}{T_2}$

    $335.6 = \dfrac{(1.5 \times 10^5)}{T_2}$

    $T_2 = \dfrac{(1.5 \times 10^5)}{335.6} = 447\ K$

Remember!
Always use temperature in Kelvin for calculations involving the gas laws.

thermometer, °C

pressure gauge

air

water

heat

Apparatus for testing the pressure law.

The gas equation

- The three gas laws are:
  - Boyle's Law: $P \propto \dfrac{1}{V}$ (temperature is kept constant)
  - Charles' Law: $V \propto T$ (pressure is kept constant)
  - The pressure law: $P \propto T$ (volume is kept constant).
- The three equations can be combined to form the gas equation:

  $\dfrac{PV}{T} = constant$

- The gas equation can be written as:

  $\dfrac{P_1 V_1}{T_1} = \dfrac{P_2 V_2}{T_2}$

  - where P_1, V_1 and T_1 are the initial conditions and P_2, V_2 and T_2 are the final conditions.
  - Temperature must be measured in Kelvin.

initial

P_1 V_1 T_1

P_2 V_2 T_2

final

A helpful way of visualising the gas equation.

- For example a cylinder contains 0.02 m³ of gas at a pressure of 12×10^5 Pa and temperature 5 °C. The gas is released into another container of volume 0.05 m³ at room temperature 25 °C. Calculate the new pressure.
  - First of all write down all the figures you have been given:

$P_1 = 12 \times 10^5$ Pa	$V_1 = 0.02$ m³	$T_1 = 273 + 5 = 278$ K
$P_2 = ?$	$V_2 = 0.05$ m³	$T_2 = 273 + 25 = 298$ K

  - Substitute the figures into the gas equation to solve for P_2

    $\dfrac{(12 \times 10^5)\, 0.02}{278} = \dfrac{P_1 \times 0.05}{298}$

    $86.3 \times 298 = 0.05 \times P_1$

    $P_1 = \dfrac{25717.4}{0.05} = 5.14 \times 10^5$ Pa

⦿ The gas equation

Higher: Explain what might happen to the temperature of a gas as it is expelled from a high-pressure cylinder into the atmosphere and occupies 5 times as much volume as it did inside the cylinder.   *AO3* [3 marks]

P3 Summary

Intensity = power of radiation/area (I = P/A)

Intensity of radiation is inversely proportional to the square of the distance from the source

Power of a lens (Dioptre) = 1 / focal length (m)

The lens equation $\frac{1}{f} = \frac{1}{v} + \frac{1}{u}$ (real is positive)

In the eye both the cornea and the lens refract light towards the retina.

Eye defects, such as long and short sight, can be corrected by wearing glasses or contact lenses, or by laser treatment.

Radiation in treatment and medicine

Total Internal Reflection (TIR) will only occur when the angle of incidence inside a more dense material is greater than a critical angle (C).

Snell's Law *refractive index* = 1/sinC.

Endoscopes use optical fibres which transmit light due to TIR.

Echoes of ultrasound (sound with frequencies greater than 20 kHz) are used to scan internal organs of the body.

The higher the frequency of X-rays the greater their energy and ionising power.

Electric current (A) = number of particles per second (1/s) × charge on each particle (C) *[I = N q]*

Kinetic energy (J) = electronic charge (C) × accelerating potential difference (V) *[KE = e V]*

X-rays are used in fluoroscopes and CAT scans.

X-rays and ECGs

An electrocardiogram (ECG) measures action potentials of the heart.

Frequency (Hz) = 1/time period (s).

A pacemaker can be used to regulate the heart action.

A pulse oximeter is used to monitor pulse rate and amount of oxygen in the blood.

Ionising radiation emitted by unstable nuclei include: alpha, beta-minus, beta-plus, gamma and neutrons.

Nuclei below the N–Z stability curve undergo beta-plus decay. (A proton becomes a positron and a neutron.)

Nuclei above the N–Z stability curve undergo beta-minus decay. (A neutron becomes an electron and a proton.)

Nuclei with large mass number (>82) undergo alpha decay. (The nucleus loses 2 protons and 2 neutrons.)

Quarks are fundamental particles. Protons and neutrons contain 3 quarks each.

Production, uses and risks of ionising radiation from radioactive sources

Ionising radiation causes cell damage, so doses are limited and workers need protection.

Palliative radiotherapy is used in the treatment of cancers and is applied either internally or externally.

Radioactive tracers are used in medical diagnoses.

PET scanners use a positron-emitting tracer.

A centripetal force acts towards the centre of a circle and accelerates the object towards the centre of the circle.

Motion of particles

In all collisions total momentum is conserved.

In elastic collisions kinetic energy is also conserved.

Particle accelerators accelerate charged particles to very high speeds using magnetic and electric fields.

If stable isotopes are bombarded with fast moving protons they will change into unstable radioactive isotopes.

When positrons collide with electrons they annihilate each other, and emit a pair of gamma rays, carrying energy $E = mc^2$, which travel in opposite directions.

As temperature increases particles move faster and make more frequent collisions.

At absolute zero (0 K or −273 °C) particles would be completely stationary. (T in Kelvin = T in Celsius + 273.)

Temperature of a gas is a measure of average kinetic energy of its particles.

Kinetic theory and gases

Pressure = force/area.

Boyle's Law states that $P \propto 1/V$.

Charles' Law states that $V \propto T$.

The pressure law states that $P \propto T$.

The gas law combines these to give $\frac{PV}{T} = constant$.

Page 4 Jupiter's moons

Higher: Galileo observed four moons orbiting Jupiter. Explain how this observation did not fit with the geocentric model of the Solar System. *AO2* [3 marks]

If moons are orbiting Jupiter, then not everything is orbiting Earth.

Answer grade: A/B. This answer is essentially correct, but it is not detailed enough to gain full marks. To improve, introduce the fact that in the geocentric model, everything in the Solar System orbited Earth.

Page 5 Refraction

Higher: Light travels more slowly in water than it does in air.

Explain why this makes it difficult to pick up an object from the bottom of a swimming pool. *AO2* [4 marks]

Light gets refracted in water so it distorts the image you see. The object at the bottom of the pool isn't where it seems to be.

Answer grade: B/C. This answer would gain 2 marks for mentioning refraction and distortion, and for accurately stating that this changes the apparent position of the object. To get full marks, however, you should say that the object appears to be closer than it really is, because the light travels more slowly in water, and your brain assumes it is travelling at the faster speed of light.

Page 6 Advantages of modern telescopes

Foundation: Early telescopes used two lenses. Modern astronomical telescopes are reflecting telescopes.

Describe **two** advantages of reflecting telescopes over refracting telescopes. *AO1* [4 marks]

The reflecting telescope uses a curved mirror instead of a lens, so it is much lighter. It is also easier to make a bigger mirror, so you can get a larger telescope.

Answer grade: C/D. This answer mentions two advantages of reflecting telescopes, but needs to include more detail. It is easier to move a lighter telescope around, and you can see more distant objects with a larger aperture telescope. Another advantage that could be mentioned is that colours refract slightly differently in a lens, so a clearer image can be obtained with a reflecting telescope.

Page 7 Thunder and lightning

Higher: During a thunderstorm you always see the lightning before you hear the thunder. Light travels so fast that the lightning is almost instantaneous, but sound travels at a speed of 340 m/s in air.

If you hear thunder half a minute after seeing lightning, how far away is the storm? *AO3* [3 marks]

$Speed = \dfrac{distance}{time}$, so $distance = speed \times time$.

$Distance = 340 \times 0.5 = 170$ *m away.*

Answer grade: C. The equation has been rearranged correctly, which would gain 1 mark, but the student has forgotten to change the time into seconds. Half a minute is 30 seconds, so the answer should be $340 \times 30 = 10\,200$ m or 10 km away.

Page 8 Electromagnetic waves

Higher: The speed of electromagnetic waves in air is about 3×10^8 m/s. Microwaves have a wavelength of 3 cm.

Use the wave equation wave speed = wave length × frequency to calculate the frequency of microwaves. *AO2* [3 marks]

Wave speed = frequency × wavelength, so frequency = speed ÷ wavelength.

$Frequency = 3 \times 108 \div 3 = 1 \times 10^8$ *Hz.*

Answer grade: B. The equation has been correctly rearranged, so one mark would be gained here. However, the units of the wavelength need to be converted to metres, not centimetres, so the calculation is incorrect. The answer should be 1×10^{10} Hz.

Page 9 Infrared waves

Foundation: All objects emit infrared waves, but the hotter the object is, the more infrared radiation is given off.

Use this information to explain how the police could use infrared radiation to find fugitives. *AO2* [3 marks]

They use infrared cameras in helicopters to find criminals on the run, because the criminals will emit infrared waves and show up in a photo.

Answer grade: D. This answer is correct, but it needs a bit more detail to gain full marks. Explain that the fugitives are warmer than their surroundings, so they will emit more infrared radiation, and that this shows up on a thermograph.

Page 10 Radioactivity

Foundation: Radioactive materials emit ionising radiation.

Explain what is meant by the term 'ionising radiation'. *AO1* [2 marks]

Ionising radiation means it turns atoms into ions.

Answer grade: E. This answer is essentially correct, but needs to explain what an ion is. For example 'when the radiation collides with an atom, it knocks off an electron to form an ion'.

Page 11 Studying the Universe

Foundation: Humans have always been fascinated with space. Ancient civilisations relied on the naked eye to study the stars. In medieval times, telescopes were used to study objects in space.

Explain **two** advantages of the technology used to study the Universe today. *AO2* [4 marks]

Nowadays we send space probes to orbit other planets, and we use the Hubble space telescope.

Answer grade: E/F. Two modern methods are identified, but no advantages are discussed. To improve, mention that the space probes can find out the elements present on other planets, and the Hubble space telescope can obtain much more detailed images of space because there is no interference from the atmosphere.

Page 12 Positioning telescopes

Higher: Explain why it is better to site optical telescopes at the top of high mountains. *AO1* [2 marks]

The air is thinner, and the light waves do not have so far to travel.

Answer grade: C/D. This answer would gain 1 mark for the first part, stating that the air is thinner. However, the student should also explain that when there is less air, the light waves will not get absorbed as much. This means that more light will arrive at the telescope, so you will be able to see further into space.

Page 13 Energy from the Sun

Higher: Explain where the Sun's energy comes from. *AO1* [4 marks]

There is a nuclear reaction inside the Sun, which gives off a lot of energy in the form of electromagnetic radiation.

Answer grade: B/C. Marks would be awarded here for correctly identifying that the energy comes from nuclear reactions in the Sun. However, to gain full marks the answer should say that the nuclear reaction is a *fusion* reaction that turns hydrogen into helium. It should also explain that some of the mass of the hydrogen is converted to energy.

Page 14 Steady State or Big Bang?

Foundation: The Steady State theory and the Big Bang theory were two opposing theories of the Universe in the 20th century.

Give one similarity between the two theories and one difference between them. *AO1* [3 marks]

They both said that the Universe is expanding. The Big Bang theory suggested that the Universe began at a certain point and has been expanding ever since.

Answer grade: C/D. This answer correctly states that both theories support the idea of an expanding Universe. However, to gain full marks, the answer should include how the Steady State theory differs from the Big Bang theory with regard to the beginning of the Universe – i.e. that according to the Steady State theory, some matter was being created all the time, not just at the beginning.

Page 15 Measuring depth

Higher: A ship uses sonar to measure the depth of water as it approaches a harbour. A short pulse of ultrasound is sent out, and the time for it to return is measured as 20 milliseconds (a millisecond is a thousandth of a second).

If the speed of ultrasound is 1500 m/s, what is the depth of water? *AO2* [2 marks]

$$distance = speed \times time = 1500 \times \frac{20}{1000} = 30 \ m$$

Answer grade: C. The time has been correctly converted into seconds, so 1 mark would be gained here. However, the student has forgotten that the distance needs to be halved, as the ultrasound has to travel to the seabed and back again. The answer should therefore be 15 m.

Page 16 S waves

Foundation: The Earth is made up of four layers – crust, mantle, outer core and inner core.

Through which layers can S waves travel? Explain your answer. *AO1* [3 marks]

S waves can travel through the crust and the mantle because they are transverse.

Answer grade: D. This answer correctly identifies the layers through which S waves can travel and also that they are transverse. However, to gain the additional mark available for this question, the student should also have explained that transverse waves cannot travel in liquids, which is why S waves cannot travel through the liquid outer core.

Page 17 Potential difference

Higher: A cell has a potential difference of 1.5 V.

What is meant by the term 'potential difference'? *AO1* [3 marks]

Potential difference is the scientific word for voltage. It pushes the electrons around the circuit.

Answer grade: B. This answer is correct and would gain 2 marks, but to achieve the additional mark available, the answer should also explain that potential difference is a measure of how much *energy* each electron gets from a battery or cell.

Page 18 Kilowatt-hours

Foundation: An oven has a power rating of 2000 W and it takes 45 minutes to cook a cake. The cost of 1 kW h of electrical energy is 22 p.

What is the cost of the electricity to cook the cake? *AO2* [2 marks]

No. of kW h = 2 × 45 = 90 kW h.

Cost = no. of kW h × 22 p = 1980 p = £19.80

Answer grade: D/E. The formulas used here are correct, but the answer is wrong. Although the power has been correctly converted to kW, the student has failed to convert 45 minutes into ¾ hour – or 0.75 hours, so the subsequent calculations are incorrect. The final answer should be 2 × 0.75 × 22 p = 33 p.

Page 19 Understanding resources

Foundation: Fuels are burnt in power stations. Fossil fuels are non-renewable energy sources, and biomass is a renewable energy source.

Explain what is meant by the terms 'non-renewable' and 'renewable'. *AO1* [2 marks]

Non-renewable means that once you have used it, it can't be used again. Renewable means it cannot be used up.

Answer grade: D. The first sentence is correct, but the answer should also include the fact that non-renewable energy sources will eventually run out. The second sentence is not strictly true. For example, biomass and wood can be used up. They are only renewable resources if crops and trees are replanted to make more. The definition of a renewable energy source is that it will not run out, not that it cannot be used up.

Page 20 Transformers

Foundation: Describe how step-up transformers are used in the National Grid. *AO1* [4 marks]

A step-up transformer is used to make voltage bigger so it can be carried across cables to people's homes.

Answer grade: E. The answer correctly identifies that a step-up transformer increases the voltage, but it does not fully explain how it is used in the National Grid. For full marks, state that each transformer from the power-station generator increases the voltage from 23 000 V to 400 000 V. Include the fact that the current in transmission cables is smaller when higher voltages are used and that this reduces heat loss in the cables, improving efficiency. Less wastage of energy also saves money.

Page 21 Calculating efficiency

Higher: An electric drill transfers every 50 J of electricity into 20 J of useful kinetic energy.

Calculate the percentage efficiency of the drill.

What forms does the wasted energy in the transfer take? *AO2* [4 marks]

$\dfrac{20}{50} = 0.4 \times 100 = 40\%$

The wasted energy is sound.

Answer grade: B. The student has used the efficiency equation correctly, and has remembered to multiply the total by 100 to reach the correct percentage. The answer also correctly identifies sound as one of the products in the energy transfer. However, for full marks, the answer should mention that heat energy is also created in the transfer.

Page 22 Staying cool

Foundation: In many Mediterranean countries, such as Greece, the houses are all painted white.

Explain why this is, using ideas about heat radiation. *AO2* [3 marks]

The white surfaces are good at reflecting the heat radiation, and poor at absorbing heat radiation.

Answer grade: D. This is scientifically correct, but needs more detail to achieve full marks. Explain that in hot countries such as those in the Mediterranean, it is useful to have a good reflector and poor absorber of heat radiation, to keep the house cool in hot weather.

Page 24 Forces and attraction

Foundation: Daisy combs her hair with a plastic comb. She can then use the plastic comb to pick up small pieces of paper.

Explain why the paper is attracted to the comb. *AO1 [3 marks]*

Because the comb and the paper have opposite charges and opposite charges attract.

> **Answer grade: E.** To achieve full marks in this question, the answer needs to include the fact that the comb has been charged by friction, as Daisy was combing her hair. It should also explain that the charged comb repels some of the electrons away from the surface of the paper, leaving the surface of the paper with a positive charge.

Page 25 Painting bicycles

Higher: A factory uses an electrostatic paint gun to paint new bicycles. Explain why the paint coats the surface of each bicycle evenly, even the back. *AO1 [2 marks]*

Because the paint is charged positive and the bicycles are charged negative. Opposite charges attract.

> **Answer grade: B.** Although this answer is correct, only one reason is given so only 1 mark would be awarded. To achieve full marks, also explain that the paint is spread out evenly. Each of the droplets of paint has a positive charge, so they repel one another, causing them to spread out evenly as they leave the nozzle.

Page 26 Calculating current

Foundation: Look at the circuit below.

What is the current at points Z and X in the circuit? *AO2 [2 marks]*

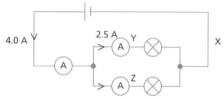

At Z the current is 2.5 A and at X it is 4.0 A.

> **Answer grade: E.** The current at X is 4.0 A, so the answer is correct here. However, the current at Z is not 2.5 A, it is 1.5 A. Notice that 2.5 A + 1.5 A = 4.0 A. The current in a circuit is always conserved.

Page 27 Variable resistors

Foundation: Explain how a circuit with a variable resistor can be used to control the brightness of a lamp in the same circuit. *AO2 [3 marks]*

When the resistance of the variable resistor is increased, the lamp is dimly lit.

> **Answer grade: D.** To improve the grade, the answer should include the value of the current. To achieve full marks, you also need to explain that the brightness of the lamp when the resistance of the variable resistor is set to its maximum value should also be explained. When the resistance of the variable resistor is set to a high value, the current in the circuit is low and the lamp is dimly lit. When the resistance of the variable resistor is set to a low value, the current in the circuit is high and the lamp shines brightly.

Page 28 Kettle calculations

Higher: Noah buys an electric kettle that supplies 2000 W of power. A full kettle of water requires 200 000 J of energy to bring it to boiling point.

a Assuming that the kettle wastes no energy, how long will it take Noah to boil a full kettle of water?

b The current in the heating element produces a heating effect. Describe how this occurs. *AO2 [3 marks]*

$$Time = \frac{energy}{power} = \frac{200\ 000\ J}{2000\ W} = 100\ s$$

The heat is produced by the movement of the electrons in the element.

> **Answer grade: B.** The calculation is correct, gaining 1 mark, but the explanation is too brief. To achieve full marks, a more detailed explanation of the heating effect must be given. For example, the movement of the electrons in the element causes energy transfer from the electrons to the atoms in the element, causing it to heat up.

Page 29 Understanding acceleration

Foundation: Syamala is training for the 1200-m race by running round a 200-m track six times. She runs at a constant speed.

Explain why she is accelerating. *AO1 [2 marks]*

Because she is changing direction.

> **Answer grade: D.** The answer is correct, but too brief. To achieve full marks, a fuller explanation is needed. Explain that velocity is a vector quantity – speed in a specific direction. As Syamala runs, her velocity is changing because her direction is constantly changing; a change in velocity is an acceleration.

Page 30 Interpreting graphs

Higher: Explain how the graph below shows that:

a The acceleration of car A is greater than the acceleration of car B.

b Car B has travelled further than car A. *AO3 [2 marks]*

Car A has a greater acceleration because its line is steeper. Car B has travelled further because its line is longer.

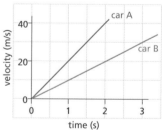

> **Answer grade: B.** Car A has greater acceleration because the line is steeper/the gradient of the line is greater. This is correct and would achieve 1 of the 2 available marks. The gradient of a velocity–time graph is equal to the acceleration. The distance each car has travelled is equal to the area under the line, so car B has travelled further than car A not simply because its line is longer, but because the area under the line for car B is greater than the area under the line for car A.

Page 31 Forces and velocity

Foundation: An aeroplane experiences the forces of gravity, upthrust, thrust from the engine and friction from the air.

Explain how it can be travelling at a constant velocity. *AO1* [3 marks]

If the forces add up to zero and there is no resultant force.

Answer grade: E. This explanation would gain 1 mark for stating that the forces add up to zero. However, with 3 marks available, more explanation is needed. If the forces on an object add up to zero, there is no resultant force and the object will either remain stationary or continue to move with constant velocity. If the upward force on the aeroplane (upthrust) equals the downward force of gravity, and the forward thrust of the engine equals the backward force of air resistance/friction/drag, then the forces add up to zero and there is no resultant force.

Page 32 Falling objects

Foundation: Kerri is a professional skydiver. When she jumps out of an aeroplane, she initially accelerates, then falls at a steady speed. Explain why. *AO1* [3 marks]

Air resistance increases until the force of gravity and the air resistance are balanced. Therefore, the resultant force is zero and she falls at a steady speed.

Answer grade: D. This explanation would achieve 2 of the 3 available marks. However, the answer has left out an important link between air resistance/friction and speed. For full marks, explain that as Kerri's speed increases, the air resistance/friction increases.

Page 33 Assessing stopping distances

Higher: Some roads have markings that indicate how far apart drivers should be from the car in front. On a particular road the markings are 30 m apart, the speed limit is 60 mph and drivers are advised to keep at least two markings apart.

Use this data and your understanding of stopping distances to evaluate whether a separation of two markings is safe in good weather. *AO3* [3 marks]

At 70 mph the total stopping distance is 93 m, so a separation of two markings is not enough.

Answer grade: C. The student has correctly identified that a separation of two markings is not enough, but the way in which he or she has reached this conclusion is not fully explained. To attain full marks, the answer should include the fact that if the distance between markings is 30 m, the distance between two markings would be 60 m. To estimate the stopping distance at 60 mph quantitatively, compare the distances for 45 mph and 70 mph. For example, at 70 mph the total stopping distance is 93 m and at 45 mph it is 46 m. At 60 mph the stopping distance would be between these values, at about 70 m. So a separation of 60 m is not a safe distance.

Page 34 Conservation of momentum

Higher: Caitlin likes to play snooker. She is learning to hit a red ball with the cue ball in such a way that the cue ball stops when it hits the red. Both the cue ball and the red ball have the same mass.

Use the principle of conservation of momentum to describe the velocity of the balls before and after the collision. *AO2* [3 marks]

Since both the balls have the same mass, the red ball will have the same velocity as the cue ball had before it hit the red ball.

Answer grade: C. This answer would only gain 1 of the available marks. For full marks, the principle of momentum should be stated – that the total momentum before the collision is equal to the total momentum after the collision – and used to explain the answer. Momentum = mass × velocity. The velocity of the red ball is zero before the collision, so the red ball has no momentum before the collision. The velocity of the cue ball is zero after the collision, so it has no momentum after the collision, as all its momentum is transferred to the red ball. Since the balls have the same mass, the velocity of the red ball after the collision is the same as the velocity of the cue ball before the collision.

Page 35 Work done

Foundation: Explain what work is done when a window cleaner climbs a ladder. *AO1* [2 marks]

The window cleaner does work climbing the ladder because he has weight.

Answer grade: E. This does not fully answer the question – it describes why the window cleaner is doing work but not what the work is. To attain full marks in this question, explain that the window cleaner is doing work against the force of gravity. Since work done = force × distance, the window cleaner will do an amount of work equal to his weight (the force) × distance (the height he climbs).

Page 36 Energy transfers

Foundation: Abeni enjoys slides at the playground.

Explain the energy transfers that occur as Abeni climbs up the slide and then slides down it. *AO2* [3 marks]

As she climbs up she gains gravitational potential energy and as she slides down she gains kinetic energy.

Answer grade: D. This answer would attain 2 of the possible 3 marks. To attain full marks, the student should expand on the answer to explain that the gravitational potential energy Abeni gains becomes kinetic energy as she slides down the slide, and that the total energy remains constant. Some of this energy will be transferred to heat through friction with the slide and the air.

Page 37 Radioactive decay

Higher: Alice has learnt at school that radioactive decay is random and cannot be predicted. Andrew argues that it must be possible to predict when radiation will occur, otherwise it would not be useable in the treatment and diagnosis of medical conditions. Who is correct? *AO2* [2 marks]

Andrew is correct. It is possible to predict when radioactive decay occurs when it is used in medicine because the samples of radioactive material have many millions of atoms in them.

Answer grade: B. While this answer is correct, to attain full marks in this question the reasons need to be explained in more detail. For example, although it is not possible to predict when an individual radioactive nucleus will decay, in a large sample of nuclei it is possible to predict roughly how many will decay in a given time.

Page 38 Controlling neutrons

Foundation: Explain how the neutrons in a nuclear power station are controlled. *AO1* [2 marks]

The control rods can be lowered into the reactor to absorb the neutrons and slow down the fission reactions.

Answer grade: D. The student has identified one way in which the neutrons in a power station are controlled, but two methods of controlling the reactions need to be given to gain both marks. As well as the control rods, the moderator in a reactor slows down the fast-moving neutrons, making them more likely to react with the uranium nuclei. Always look at the number of marks available to get an idea of how much information you should give in your answers.

Page 39 The future of fusion

Foundation: Scientists hope that fusion reactors will soon be able to produce energy for human consumption.

a What fuel will be used in these reactors?

b Explain how nuclear fusion occurs.
AO1/AO2 [3 marks]

Hydrogen is the fuel for these reactors. Hydrogen nuclei join together to produce helium.

Answer grade: D. This answer correctly identifies hydrogen as the fuel, but the answer to part **b** of the question is not complete. To push this up to a grade C, include the fact that the hydrogen nuclei collide at high speeds in order to fuse/join together to form helium nuclei.

Page 40 Differences in background radiation

Foundation: Joe and Fred are pen-pals who live in different parts of the UK. They have both been learning about background radiation at school and have discovered that the level of background radiation where they live is different.

Explain why this is. *AO1* [3 marks]

Because the rocks that give off radiation are different in each location.

Answer grade: E. This answer is too superficial to earn more than 1 mark. To achieve full marks, explain that the rock granite contains uranium. Uranium decays to produce radioactive radon gas. Different parts of the UK have different amounts of granite in the landscape, so the amount of background radiation varies.

Page 41 Using radiation

Higher: Which type of radiation would be most suitable for use in a machine that controls the thickness of thin aluminium sheeting? Give the reasons for your answer. *AO2* [3 marks]

Beta radiation would be most useful because it can pass through the aluminium. Alpha radiation could not be used because it cannot pass through aluminium.

Answer grade: B. This answer mentions both alpha and beta radiation. However, the student has overlooked gamma radiation. It is correct that beta radiation can pass through a thin sheet of aluminium, but gamma radiation can also penetrate aluminium. The advantage of beta radiation is that it cannot penetrate through a thick sheet of aluminium, so the amount of beta radiation passing through the sheet would change as the thickness of the sheet changes. It would not be possible to measure any change in the amount of gamma radiation passing through a sheet of aluminium.

Page 42 Measuring radioactive decay

Higher: The graph below shows the count rate of a radioactive sample over a period of 40 days. Using the graph, estimate the half-life of the sample. *AO3* [2 marks]

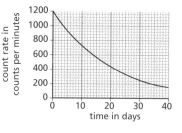

13.5 days.

Answer grade: B. The answer is correct, but in order to achieve both of the available marks, some working should be shown. For example, reference lines could be drawn on the graph. Remember, you should always show your working out. Sometimes marks can be earned for correct parts of working, even if the final calculation is wrong.

Page 44 Intensity of light

Higher: A street light has a much more powerful bulb than a hand-held battery-powered torch. Explain why the smaller torch can light up the pavement almost as well as the street light. *AO2* [3 marks]

The street light is much further away so the light from it has got more spread out. You are holding the torch closer to where you want to walk and it has a narrow beam.

Answer grade: C. This contains the basic idea, but needs to include a more scientific explanation. For example, the intensity of light is inversely proportional to the distance away from the bulb. You could also add that the narrow beam of the torch has a much smaller area, so although the light source is dimmer the light does not spread out much.

Page 45 Power of a lens

Higher: A diverging lens has a focal length of 40 cm. Calculate the optical power of the diverging lens. *AO1* [3 marks]

40 cm = 0.4 m.

$Power = \dfrac{1}{0.4} = 2.5$

Answer grade: C. This student has remembered to convert centimetres to metres, but has forgotten that a diverging lens has a negative optical power. Also the unit of optical power should be included. The answer should be −2.5 D.

Page 46 Images formed using converging lenses

Foundation: The top ray diagram on this page shows how an image is formed by a converging lens. If the object stays in the same place, but you replace the lens with one which has a shorter focal length, explain what will happen to the image. *AO2* [3 marks]

It will get bigger, because the light rays will bend more.

Answer grade: G. The light rays do bend more, but this will mean that the rays of light will cross at a position closer to the lens and closer to the principal axis. This means that the image will be smaller.

Page 47 Short sight

Higher: If a short-sighted person looks at a very distant object, the image is blurred. Explain why. *AO1* [3 marks]

When a short-sighted person looks at things far away the image is formed in front of the retina, so it is not focused properly on the retina, so it is blurred.

Answer grade: C/B. This answer needs more detailed explanation. It should say that either the lens cannot become thin enough, or the eyeball is too long.

Page 48 Total internal reflection

Foundation: Explain the difference between internal reflection and total internal reflection. *AO1* [3 marks]

Internal reflection means light is reflected inside. Total internal reflection is when all the light is reflected inside the glass block and nothing is refracted.

Answer grade: F. This is a good start to the answer, but needs to be developed further by saying that some light is always internally reflected as well as refracted, and that total internal reflection only occurs when the angle of incidence is greater than a critical angle.

Page 49 Ultrasound

Higher: During an ultrasound scan the radiographer will use a jelly lubricant between the transducer and the skin. Using your knowledge of sound waves, explain why this is done. *AO2* [3 marks]

The jelly lubricant makes the transducer slide over the skin without causing pain to the patient. It also perhaps helps the ultrasound pass through the skin.

Answer grade: D. This answer is partially correct, but does not use any knowledge of how sound waves travel, so cannot score very many marks. Sound waves travel faster in solids and liquids than in air, so the sound waves will travel through the jelly into the skin without much reflection at the air/skin boundary. This means that the intensity of the ultrasound used does not need to be so high.

Page 50 Evacuated tubes

Foundation: Explain what is meant by the term 'thermionic emission'. *AO1* [3 marks]

Electrons are emitted from the cathode by thermionic emission.

Answer grade: F. What the candidate has written is correct, but does not really explain the process. You need to say that the cathode is heated by an electric current flowing in the filament, and that this gives electrons sufficient energy to escape.

Page 51 Absorption of X-rays

Higher: Explain what is meant by an 'inverse square law' for the absorption of X-rays. *AO1* [2 marks]

If you double the distance away from an X-ray source the energy of the X-rays will reduce by a quarter.

Answer grade: B. The candidate has defined the inverse square law mathematically for one example, but has used the incorrect term. It is not the energy which is reduced, but the intensity, which will be reduced by a quarter. You also need to specifically say that the intensity will decrease in proportion to the square of the distance from the source.

Page 52 Heart action and pacemakers

Higher: Explain when a doctor might implant a pacemaker in a patient's heart. *AO1* [2 marks]

Someone might need a pacemaker if their heart does not have a regular heartbeat.

> **Answer grade: C.** This is basically correct, but pacemakers will only be implanted in the most severe cases, and after the patient has tried the use of drugs with no success.

Page 53 Pulse rates

Foundation: Why is it important to monitor a patient using pulse oximetry when he or she is under anaesthetic? *AO2* [2 marks]

Doctors need to know that the patient is still alive, so they need to check that there is still a pulse rate.

> **Answer grade: F.** This only mentions one reason, so you also need to say that the amount of oxygen in the blood also needs to be monitored. If the amount of oxygen in the blood gets too low, the anaesthetist can use an oxygen mask so the patient breathes in more oxygen.

Page 54 The nuclear atom

Foundation: Carbon 14 is a radioactive isotope which is written $^{14}_{6}C$. How many protons and neutrons are present in the nucleus? *AO1* [2 marks]

There are 14 neutrons and 6 protons in the nucleus.

> **Answer grade: F.** This is the correct number of protons, but the number of neutrons is 8. The top number in the notation refers to the total number of particles in the nucleus, so to work out the number of neutrons you must find the difference between the two numbers.

Page 55 Beta-minus decay

Higher: A beta-minus particle is a fast moving electron. Explain how an electron can be emitted from the nucleus, which only contains protons and neutrons. *AO1* [2 marks]

The nucleus loses an electron, and has to gain a proton to maintain the same overall charge.

> **Answer grade: D.** This answer does not mention the fact that a neutron splits into a proton and an electron. The electron is emitted as a beta-minus particle, but the proton stays in the nucleus.

Page 56 Type of decay

Higher: Silicon-27 is an unstable isotope of silicon. Silicon has an atomic number of 14. Use your knowledge about stability of nuclei to work out which type of beta decay it will undergo. Explain your answer. [3 marks *AO3*]

If silicon had equal numbers of protons and neutrons it would be $^{28}_{14}Si$. This isotope has fewer neutrons than this, so it needs to lose a neutron to become stable. It will undergo beta-minus as a neutron will change to a proton and a electron.

> **Answer grade: C.** This is actually the incorrect type of decay, but the first sentence is a good start. However, the candidate then got confused by saying that the neutron-poor nucleus needs to lose another neutron. It actually has too many protons, so it undergoes beta-plus decay, as a proton will change to a neutron and a positron.

Page 57 Fundamental particles

Higher: Explain why electrons are said to be fundamental particles, but protons are not. *AO1* [2 marks]

Electrons are fundamental particles because they are not made up of quarks; protons are made up of three quarks.

> **Answer grade: A/B.** This is basically correct, but you need to define fundamental particles as those particles which cannot be subdivided.

Page 58 Damage to cells

Higher: Explain why blood cells and hair follicles are more susceptible to damage by ionising radiation. *AO2* [2 marks]

Because they are rapidly dividing.

> **Answer grade: C.** This answer is not detailed enough for 2 marks. These cells are rapidly dividing, but any abnormality in one cell caused by the ionising radiation will be replicated each time the cell divides, so a large number of abnormal cells will be produced quickly.

Page 59 Palliative radiotherapy

Foundation: Palliative radiotherapy is often given to cancer sufferers. Describe one benefit and one drawback of this type of therapy. *AO1* [2 marks]

It cures cancer, but it makes you feel sick.

> **Answer grade: F.** The drawback is correct, but palliative radiotherapy does not cure cancer, it only shrinks it, or slows down its growth.

Page 60 Radioactive tracers

Foundation: There are many radioactive isotopes of iodine. Iodine-131 is a beta emitter with a half-life of about 8 days. Iodine-108 is an alpha emitter with a half-life of 36 ms. Explain why Iodine-108 cannot be used as a radioactive tracer. *AO3* [3 marks]

Iodine-108's half-life is too short. The doctors would not be able to see the results within the time of the half-life. Iodine-131 has a half-life of a few days so it is okay.

> **Answer grade: D.** This is correct, but iodine-108 is also an alpha emitter. Alpha particles are too ionising to be used inside the body. You need to mention both reasons to get full marks on this question.

Page 61 Particle accelerators

Foundation: Particle accelerators can either be linear or circular. Describe one advantage of circular particle accelerators over linear ones. *AO2* [2 marks]

In a circular particle accelerator the particles go round and round in circles getting faster each time, but in a linear accelerator they just go one way along it.

> **Answer grade: F.** This answer goes part way to answering the question, but needs to explain that a circular accelerator will take up less space than a linear accelerator. Because the particles move in a circular path, the diameter of the accelerator will be less than the length of a linear one.

Page 62 Cyclotrons

Higher: In a cyclotron similar to the one in the diagram above, the particles describe a spiral path. Explain how the magnitude and direction of the particle's velocity changes during this spiral path. *AO2* [3 marks]

When the particle is going along its semi-circular path there is a centripetal force towards the centre of the circle and it goes at constant speed. When it goes from one dee to the next its speed increases but the direction stays the same until it gets into the next dee, when it makes another larger semi-circular path.

> **Answer grade: C.** The first sentence does not really answer the question; whilst the particle is travelling in a semi-circular path the magnitude of the velocity is constant, but its direction is changing all the time. Between the two dees the magnitude of the velocity increases, but the direction is unchanged. The next semi-circular path will be at a greater radius as the magnitude of the velocity is now larger than before. This creates the spiral path.

Page 63 Matter, antimatter and annihilation

Foundation: During a collision between an electron and a positron, explain what physical quantities are conserved. *AO1* [3 marks]

Charge, momentum and energy are conserved.

> **Answer grade: F.** The question did not just say to state what was conserved during the annihilation. You have been asked to explain, so the answer requires more detail. For example: charge is conserved because the net charge of the two particles is zero and the gamma rays given off have no charge; momentum is conserved because the two gamma rays go off in opposite directions and it is mass-energy which is conserved, because the total mass of the two particles is converted to energy.

Page 64 Inelastic collisions

Higher: An object of mass 150 g travels in a straight line with velocity 4 m/s and collides with another object with mass 100 g which is moving in the opposite direction with a velocity 3 m/s. The two objects stick together. Calculate their common velocity. *AO2* [3 marks]

Momentum before collision = $(0.15 \times 4) + (0.1 \times 3) = 0.9$

New velocity = momentum/total mass = $\dfrac{0.9}{0.25} = 3.6$ m/s

> **Answer grade: B.** The working is clearly laid out, but the candidate has forgotten that the two objects are initially travelling in opposite directions, so the velocities should have opposite signs. The initial total momentum = 0.3 kg m/s, so the final velocity = $\dfrac{0.3}{0.25} = 1.2$ m/s.

Page 65 Kinetic theory of matter

Foundation: Describe the arrangement and movement of particles in a liquid as its temperature increases. *AO1* [3 marks]

The particles are close together and they are moving around each other.

> **Answer grade: E.** This is a basic description of the arrangement and movement of particles in a liquid, but does not describe what happens as the temperature increases. You need to say that as the temperature increases the particles move faster. You should also state that although the particles are close together, they are not in a regular array.

Page 66 Boyle's Law

Higher: When you decrease the volume of a fixed mass of gas at a constant temperature, its pressure increases. Explain this using ideas about the kinetic theory of gases. *AO2* [3 marks]

In the gas all the particles are moving around fast and colliding with each other. As you decrease the volume they will have more collisions, and the force exerted in collisions makes the pressure.

Answer grade: C/B. This answer is along the right lines, but you need to clarify that you are talking about collisions with the walls of the container. The particles do also collide with each other, but it is the collisions with the walls that causes the pressure. As the volume decreases, if the particles are moving at a constant average speed (at constant temperature), then they will travel across the smaller width of the container in a shorter time, so will make more frequent collisions with the wall.

Page 67 The gas equation

Higher: Explain what might happen to the temperature of a gas as it is expelled from a high-pressure cylinder into the atmosphere and occupies 5 times as much volume as it did inside the cylinder. *AO3* [3 marks]

As the gas escapes its volume increases, and because $V \propto T$, the temperature must increase.

Answer grade: C. This is not necessarily correct, as Charles' Law is only true for constant pressure. The pressure of the gas is reducing by a large amount as well as the volume increasing. If the pressure decreases by the same amount as the volume increases then there will be no change in temperature. However, if the pressure decreases by more than 5 times, then the temperature will decrease.

How Science Works

Data, evidence, theories and explanations

As part of your Science and Additional Science assessment, you will need to show that you have an understanding of the scientific process – How Science Works.

This involves examining how scientific data is collected and analysed. You will need to evaluate the data by providing evidence to test ideas and develop theories. Some explanations are developed using scientific theories, models and ideas. You should be aware that there are some questions that science cannot answer and some that science cannot address.

Practical and enquiry skills

You should be able to devise a plan that will answer a scientific question or solve a scientific problem. In doing so, you will need to collect data from both primary and secondary sources. Primary data will come from your own findings – often from an experimental procedure or investigation. While working with primary data, you will need to show that you can work safely and accurately, not only on your own but also with others.

Secondary data is found by research, often using ICT – but do not forget books, journals, magazines and newspapers are also sources. The data you collect will need to be evaluated for its validity and reliability as evidence.

Communication skills

You should be able to present your information in an appropriate, scientific manner. This may involve the use of mathematical language as well as using the correct scientific terminology and conventions. You should be able to develop an argument and come to a conclusion based on recall and analysis of scientific information. It is important to use both quantitative and qualitative arguments.

Applications and implications of science

Many of today's scientific and technological developments have both benefits and risks. The decisions that scientists make will almost certainly raise ethical, environmental, social or economic questions. Scientific ideas and explanations change as time passes and the standards and values of society change. It is the job of scientists to validate these changing ideas.

How science ideas change

From the information you have learnt, you will know that science is a process of developing, then testing theories and models. Scientists have been carrying out this work for many centuries and it is the results of their ideas and trials that has provided us with the knowledge we have today.

However, in the process of developing this knowledge, many ideas were put forward that seem quite absurd to us today.

> In 1692, the British astronomer Edmund Halley (after whom Halley's Comet was named) suggested that the Earth consisted of four concentric spheres. He was trying to explain the magnetic field that surrounds the Earth and suggested that there was a shell of about 500 miles thick, two inner concentric shells and an inner core. Halley believed that these shells were separated by atmospheres, and each shell had magnetic poles with the spheres rotating at different speeds. The theory was an attempt to explain why unusual compass readings occurred. He also believed that each of these inner spheres, which was constantly lit by a luminous atmosphere, supported life.

Reliability of information

It is important to be able to spot when data or information is presented accurately and just because you see something online or in a newspaper, does not mean that it is accurate or true.

Think about what is wrong in this example from an online shopping catalogue. Look at the answer at the bottom of the page to check that your observations are correct.

> ### From box to air in under two minutes!
>
> Simply unroll the airship and, as the black surface attracts heat, watch it magically inflate.
>
> Seal one end with the cord provided and fly your 8-metre, sausage-shaped kite.
>
> - Good for all year round use.
> - Folds away into box provided.
> - A unique product – not for the faint hearted.
> - Educational as well as fun!
>
> Once the airship is filled with air, it is warmed by the heat of the sun.
>
> The warm air inside the airship makes it float, like a full-sized hot-air balloon.

Answer

Black absorbs heat, it does not attract it.

Glossary

A

acceleration the rate of change of the velocity of an object 29, 30, 31, 32, 34

activity the activity of a source is the rate of decay of nuclei 42

alpha particle a helium nucleus emitted from an unstable nucleus 10, 37, 38, 41, 42

alternating current a current that repeatedly changes direction 20

ammeter meter used in an electric circuit for measuring current 17, 18, 26, 28

amperes (amps) units used to measure electrical current 17, 26, 28

amplitude the maximum displacement of a wave measured from the mean position 7

anode positively charged electrode 50

antimatter matter made up of anti-particles, such as positrons 63

atmosphere the layer of gases surrounding a planet 4, 12, 40

atom the basic 'building block' of an element which cannot be chemically broken down 10, 22, 24, 37

atomic number the number of protons inside a nucleus 54

B

becquerel the unit for activity: one becquerel is equal to one nucleus decaying per second 42

beta particle an electron emitted from the inside of an unstable nucleus 37, 38, 41, 42

Big Bang an explosion some 14 billion years ago that created both space and time 14

Big Bang theory a theory that proposes the creation of the Universe from the Big Bang 14

biomass the amount of organic material of an organism (usually measured as dry mass); waste wood and other natural material which are burned in power stations 19

black dwarf the final stage of a white dwarf, when it has lost all its energy 13

black hole an extremely dense core of a supermassive star left behind after the supernova stage; light cannot escape its strong gravitational pull 13

braking distance the distance travelled by a car while the brakes are applied and the car comes to a stop 33

C

cathode negatively charged electrode 50

centripetal force the resultant force acting at right angles to the velocity of an object that gives rise to circular motion 62

chain reaction a process in which an enormous amount of energy is produced when neutrons from previous fission reactions go on to produce further uncontrolled fission reactions 38

charge a physical property of particles which causes them to experience a force when near other electrically charged particles 17, 24, 25, 26

chemical energy energy available from atoms when electron bonds are broken 17, 21, 36

cold fusion an invalidated theory that proposed nuclear fusion occurring at room temperature 39

compressions regions where particles are pushed together and create a region of higher pressure in a sound wave 7

control rods material used to absorb the neutrons in a nuclear reactor in order to produce a controlled chain reaction 38

conventional current the flow of positive charges 17

converging lens a lens that focuses parallel rays of light to a point 5, 6

coolant gas or liquid used to remove thermal energy from a nuclear reactor 38

cosmic microwave background (CMB) radiation the 'left-over' radiation from the Big Bang – radiation coming very faintly from all directions in space 14, 30

coulomb the unit for charge 17, 26

critical angle the angle of incidence in a denser medium that gives an angle of refraction equal to 90° 48

critical mass the minimum mass of fissile material that can sustain a chain reaction 38

current the rate of flow of charge 17, 18, 20, 26, 27, 28

cyclotron a particle accelerator used to produce radioactive isotopes used in PET scans 60

D

daughter nuclei the nuclei produced in a fission reaction 28, 29, 30

deceleration negative acceleration 29, 30

density the density of a substance is found by dividing its mass by its volume 5

diminished an image that is smaller than the object 6

diode a device made from semiconductor material that conducts in one direction only 27

dioptre a unit for the optical power of a lens 45

direct current an electric current that flows in one direction only 20, 26

displacement distance moved in a specific direction 7, 29

diverging lens a lens that makes parallel rays of light spread out rather than focus to a point 5, 6

Doppler effect the change in wavelength or frequency of a wave as a result of relative motion between the source and an observer 12, 14

E

earthing a method used for ensuring the safe discharge of charges to the Earth 25

echolocation a technique similar to sonar used by some animals to navigate and find their prey 5, 15

efficiency the proportion of the input energy that is transferred to useful form, calculated using the equation: efficiency = (useful energy transferred by the device / total energy supplied to the device) x 100% 21

elastic collision a collision in which momentum and kinetic energy are both conserved 64

electric current when electricity flows through a material we say that an electric current flows 17, 21, 26, 28

electric force field a region where electric charges experience a force 25

electrical power the rate of energy transfer 18, 22, 28

electromagnetic spectrum electromagnetic waves ordered according to wavelength and frequency, ranging from low-frequency radio waves to high-frequency gamma rays 8, 9, 10, 11, 12

electrons tiny negatively charged particles within an atom that orbit the nucleus – responsible for current in electrical circuits 10, 17, 24, 25, 26, 28, 37, 41

electrostatic forces the very strong forces between positive and negative ions in an ionic substance 24, 25

element a substance made out of only one type of atom 37

endoscope an instrument used by doctors to look inside the body 48

exponential decay a graph in which the quantity halves after a given interval of time 42

extraterrestrial a term used to describe things 'beyond Earth' 11

eyepiece the lens at the end of a telescope that you look through 6

F

far point the furthest point the eye can see clearly 47

filament lamp a lamp that emits light when its thin metal filament gets very hot 27

Glossary

fission reactions the splitting of a nucleus when it absorbs a neutron 38, 39

fixed resistor a resistor that can only resist a specific amount of current 27

focal length the distance between the centre of the lens and the focal point 5, 6, 45

force of gravity an attractive force between all particles that have mass 11,

fossil fuels fuel (coal, natural gas, oil) formed from the compressed remains of plants and other organisms that lived long ago 19, 42

fossils the preserved remains of organisms that lived long ago 16

free-body force diagram a diagram showing all the forces acting on an object 31

frequency the number of vibrations per second or number of complete waves passing a set point per second 7, 8, 10, 12, 15, 20

friction energy losses caused by two or more objects rubbing against each other 16, 24, 33

fuel rods rods containing nuclear fuel for a fission reactor 38

fuse a thin wire used in an electrical circuit as a safety device 18

fusion reactions reactions in which lighter nuclei join together (fuse) and produce energy 13, 38, 39

G

galaxy a collection of billions of stars held together by the force of gravity 11, 14

gamma rays electromagnetic waves of short wavelength emitted from unstable nuclei 10, 37, 41, 42

generator a device used for producing electrical energy by moving wires through a magnetic field 20

geocentric model Earth-centred model of the Solar System 4

gradient a quantity determined by dividing the change in y by the change in x; it is the slope of a line 30

gravitational field strength gravitational force acting on an object per unit mass 32, 36

gravitational potential energy the energy associated with the position of an object in the Earth's gravitational field 21, 35, 36

greenhouse effect a process in which the atmosphere is warmed up by infrared radiation; it then re-radiates some of the infrared radiation back towards the Earth's surface, which warms the surface 19

greenhouse gases gases in the atmosphere whose absorption of infrared solar radiation is responsible for the greenhouse effect, e.g. carbon dioxide, methane and water vapour 19

H

haemoglobin a protein molecule in the red blood cells that carries oxygen from the lungs to the body's tissues 53

half-life the half-life of an isotope is the average time taken for half of the undecayed nuclei in the sample to decay 40, 41, 42

heliocentric model Sun-centred model of the Solar System 4

I

induced a term used to mean 'created' 20, 24

inelastic collision a collision in which momentum is conserved but kinetic energy is not; some of the kinetic energy is transformed to other forms such as heat and sound 64

infrared radiation part of the electromagnetic spectrum, thermal energy 9, 19, 22

infrared waves non-ionising waves with a wavelength longer than red light that are radiant heat 4, 6, 8, 9, 10

infrasound sound with frequencies less than 20 Hz 15

insecticide chemicals used to kill insects 25

insulators materials that are poor electrical conductors 24, 25, 37

intensity the radiant power per unit area 28

ion an atom with an electrical charge (can be positive or negative) 10, 24, 25, 37

ionisation a process in which radiation transfers some or all of its energy to liberate an electron from an atom 10, 37, 41

ionosphere a region of charged particles around the Earth that reflects radio waves 5, 9

isotopes (1) nuclei of atoms with the same number of protons but a different number of neutrons; (2) atoms with the same number of protons but different numbers of neutrons 37, 39, 40, 42, 54

J

joule the unit of work done and energy: one joule is the work done when a force of 1 N moves a distance of 1 m in the direction of the force 35, 36

K

kilowatt-hour the energy used by an appliance of power 1 kW used for 1 hour – it is a unit of energy used by electrical suppliers 28

kinetic energy the energy that moving objects have 21, 22, 36, 38

L

lander a robotic space probe sent to planets and moons to carry out soil analysis 21

lattice a criss-cross structure 28

life cycle a term used to describe the journey of a star in time 13

light-dependent resistor (LDR) a device in an electric circuit whose resistance falls as the light falling on it increases 28

light-year the distance travelled by light in a vacuum in one year 11

longitudinal waves waves with vibrations parallel to the direction in which they travel 7, 16

M

magma hot molten rock found in the mantle, below the Earth's surface 16

magnified an image that is larger than the object 6

main sequence star an average star, just like our Sun 13

mass the amount of matter inside an object, measured in kilograms 32, 34, 36

mass–energy equation Einstein's equation $E = mc^2$, which links mass with the energy of a system 63

mass number the total number of neutrons and protons within the nucleus of an atom (same as the nucleon number) 37, 54

microorganisms single-celled organisms that are only just visible in the light microscope 41

moderator material used to slow down the fast-moving neutrons in a nuclear reactor 38

momentum a quantity calculated by multiplying the mass of an object by its velocity 34, 63

N

National Grid the network of pylons, high-voltage cables and transformers that carries electricity from power stations across the country 20

near point the closest distance the eye can focus an object 47

nebula a cloud of dust and gas from which stars form 13

neutron star a very dense core of a massive star left behind after the supernova stage 13

neutrons small particles that do not have a charge, found in the nucleus of an atom 24, 37, 38, 39

Glossary

Newton's second law a law expressed by the equation: force = mass x acceleration 32

non-renewable resources resources which are being used up more quickly than they can be replaced, e.g. fossil fuels; they will eventually run out 19, 42

nuclear fusion the fusing together of hydrogen nuclei to produce helium nuclei 13, 38, 39

nucleon number the total number of neutrons and protons within the nucleus of an atom (same as the mass number) 37, 54

nucleons a term used to refer to either protons or neutrons 37

nucleus the central core of an atom, which contains protons and neutrons and has a positive charge 10, 24, 37, 38, 39

O

objective lens the lens at the front of a telescope 6

optical fibres thin and flexible tubes of transparent material for transmitting light from one end to another 48

optical power a quantity found using '1/focal length of the lens in metres' 45

P

pacemaker a device implanted into a patient to regulate heartbeats 52

parallel when components are connected across each other in a circuit 17, 26

payback time the number of years it takes to get back the cost of an energy-saving method 18

peak the uppermost point of a wave 7

photography the process of producing permanent images 4, 8, 9

potential difference another term for voltage (a measure of the energy carried by an electric charge) 17, 18, 26, 27, 28, 41

power the rate of work done or the rate of energy transfer 18, 28, 35

pressure law the pressure exerted by a fixed mass of gas, kept at constant volume, is directly proportional to its temperature on the Kelvin scale 67

principal focus the point at which rays of light parallel to the principal axis of a converging lens converge, or the point from which rays parallel to the principal axis of a diverging lens appear to come 45

principle of conservation of energy energy cannot be created or destroyed, it can simply be transferred from one form to another 21, 36

principle of conservation of momentum for a system of colliding objects, where there are no external forces, the total momentum before and after the collision remains the same 34, 64

prism a block of glass used to split white light into a visible spectrum 8, 12

proton number the number of protons inside the nucleus of an atom (same as the atomic number) 37, 54

protons small positively charged particles found in the nucleus of an atom 24, 37

R

radiopharmaceutical a substance produced by tagging radioactive isotopes to natural chemicals such as glucose and water 60

radiotherapy a technique that uses gamma rays to kill cancer cells in the body 10, 41

radon a colourless, odourless and radioactive gas originating from rocks such as granite 40

rarefactions regions where particles are pulled apart and create regions of low pressure in sound waves 7

real image an image formed on the other side of the lens to the object – a real image can be formed on a screen 6

reflecting telescope a telescope with a concave mirror, a flat mirror and an eyepiece 6

reflection when a wave is bounced off a surface 5, 9

refracting telescope a telescope that uses two convex lenses to collect and focus light 6

refraction the bending of a wave caused by the change in its speed – when a light ray travelling though air enters a glass block it changes direction 5, 6, 9, 12, 16

renewable resources energy resources that can be replenished at the same rate that they are used up, e.g. biofuels – they will not run out 19

resistance an electrical quantity determined by dividing potential difference by current 26, 27, 28

resources the raw materials taken from the environment and used to run industry, homes and transport, and to manufacture goods 19

rheostat a variable resistor 27

S

Sankey diagram a diagram showing the transfer of energy to different forms 21

scalar a quantity that only has size or magnitude 29, 32

secondary coil the output coil of a transformer 20

seismic waves shock waves from earthquakes 7, 16

seismometer an earthquake-detecting instrument 7, 16

semiconductor a group of materials with electrical conductivity properties between metals and insulators 28

series when components are connected end-to-end in a circuit 17, 26

Snell's law an equation that relates the angle of incidence i in a vacuum (or air), the angle of refraction r in a medium and the refractive index n of the medium; $\sin i/\sin r = n$ 48

Solar System the Sun and all the objects orbiting it (planets, asteroids, comets, etc.) 4, 11

sonar a technique used by ships to determine the depth of water: it stands for Sound Navigation And Ranging 15

spectrometer a device used to analyse light from various sources 12

speed how fast an object travels, calculated using the equation: speed (metres per second) = distance / time 5, 7, 8, 15, 16, 29, 30, 31, 32, 33, 35, 36

Steady State theory a theory that proposes a Universe in which matter is created from empty space to keep its density the same 14

step-down transformer a device used to change the voltage of an a.c. supply to a lower voltage 20

step-up transformer a device used to change the voltage of an a.c. supply to a higher voltage 20

sterilisation a technique used to kill bacteria by exposure to radiation 41

strong nuclear force an attractive force between all neutrons and protons 56

super red giant a huge expanded star larger than a red giant 13

supernova an exploding star 13

T

tangent a line drawn to a curve to determine the gradient of a curve at a point 30

target the anode in an X-ray tube 50

tectonic plates the several solid parts of the Earth's crust 86

telescope a device using lenses (or mirrors) to magnify distant objects 4, 5, 6, 11, 12, 14

terminal velocity the constant velocity of a falling object when the net force acting on it is zero 32

Glossary

thermionic emission emission of electrons from the surface of a heated metal 50

thermistor a sensor in an electric circuit that detects temperature 28

thermograph an image produced using infrared waves 19

thinking distance the distance travelled by a car as the driver reacts to apply the brakes 33

total internal reflection a phenomenon where 100% of the light is reflected back into a material, when the ray hits the glass/air boundary at an angle that is greater than the critical angle 9, 48

tracer a radioactive material injected into a patient for locating cancer or diagnosing a function of the body 10, 41

transducer a device for converting electrical energy to sound and vice versa 49

transformer a device that converts the voltage of an a.c. supply to another voltage 20

transverse waves waves with vibrations at right angles to the direction in which the wave is travelling 7, 8, 16

trough the lowest point of a wave 7

U

ultrasound sound with frequencies greater than 20 kHz – too high for detection by human ears 5, 15

ultraviolet waves electromagnetic waves with a wavelength shorter than violet (blue) light 4, 6, 8, 9, 10

Universe the whole of space containing all the galaxies 4, 11, 14

V

vacuum empty space that has no particles 8, 32

validated to establish the soundness of, or to corroborate, evidence 39

vector a quantity that has both magnitude and direction 29, 31, 32, 34

velocity how fast an object is travelling in a certain direction: velocity = displacement / time 29, 30, 31, 32, 35, 36

virtual image an image formed on the same side of the lens as the object – a virtual image can be seen by looking though the lens, it cannot be projected onto a screen 6

volt the unit of voltage 17

voltage the energy transferred per unit charge – a measure of the energy carried by electric charge (also called the potential difference) 17, 20, 26, 27

voltmeter a device used to measure the voltage across a component 17, 18, 26, 28

W

watt the unit for power 18, 35

wavelength the distance between neighbouring wave peaks or wave troughs 7, 8, 9, 12, 14

weight the gravitational force acting on an object, measured in newtons 31, 32, 35, 36

white dwarf a hot and dense core of a star (such as our Sun) left behind after the red giant stage 13

work done the product of force and distance moved in the direction of the force 35, 36

X

X-rays electromagnetic waves with very short wavelength of the order of 0.000 000 001 m 4, 8, 10, 11, 12

Exam tips

The key to successful revision is finding the method that suits you best. There is no right or wrong way to do it.

Before you begin, it is important to plan your revision carefully. If you have allocated enough time in advance, you can walk into the exam with confidence, knowing that you are fully prepared.

Start well before the date of the exam, not the day before!

It is worth preparing a revision timetable and trying to stick to it. Use it during the lead up to the exams and between each exam. Make sure you plan some time off too.

Different people revise in different ways and you will soon discover what works best for you.

> **Remember!**
>
> There is a difference between *learning* and *revising*.
>
> When you revise, you are looking again at something you have already learned. Revising is a process that helps you to remember this information more clearly.
>
> Learning is about finding out and understanding new information.

Some general points to think about when revising

- Find a quiet and comfortable space at home where you won't be disturbed. You will find you achieve more if the room is ventilated and has plenty of light.

- Take regular breaks. Some evidence suggests that revision is most effective when tackled in 30 to 40 minute slots. If you get bogged down at any point, take a break and go back to it later when you are feeling fresh. Try not to revise when you're feeling tired. If you do feel tired, take a break.

- Use your school notes, textbook and this Revision guide.

- Spend some time working through past papers to familiarise yourself with the exam format.

- Produce your own summaries of each module and then look at the summaries in this Revision guide at the end of each module.

- Draw mind maps covering the key information on each topic or module.

- Review the Grade booster checklists on pages 153–155.

- Set up revision cards containing condensed versions of your notes.

- Prioritise your revision of topics. You may want to leave more time to revise the topics you find most difficult.

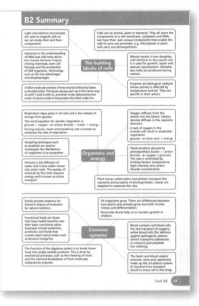

Workbook

The **Workbook** (pages 87–152) allows you to work at your own pace on some typical exam-style questions. You will find that the actual GCSE questions are more likely to test knowledge and understanding across topics. However, the aim of the Revision guide and Workbook is to guide you through each topic so that you can identify your areas of strength and weakness.

The Workbook also contains example questions that require longer answers (**Extended response questions**). You will find one question that is similar to these in each section of your written exam papers. The quality of your written communication will be assessed when you answer these questions in the exam, so practise writing longer answers, using sentences. The **Answers** to all the questions in the Workbook can be cut out for flexible practice and can be found on pages 156–166.

At the end of the Workbook there is a series of **Grade booster checklists** that you can use to tick off the topics when you are confident about them and understand certain key ideas. These Grade boosters give you an idea of the grade at which you are currently working.

Notes

Collins
Workbook

NEW GCSE

Physics

Foundation and Higher

for Edexcel

Authors: **Sarah Mansel**
Caroline Reynolds

Revision Guide +
Exam Practice Workbook

The Solar System

1 The ancient Greeks used a geocentric model of the Solar System.

a What is meant by the term 'geocentric'?

....

.... **[2 marks]**

b Name the five planets in the geocentric model.

....

.... **[3 marks]**

c Why are there only five planets in the geocentric model when there are more in the modern model of the Solar System?

....

.... **[2 marks]**

d Describe an observation that the ancient Greeks found difficult to explain using the geocentric model, but which was easier to explain using the heliocentric model.

....

....

....

.... **[3 marks]**

2 Why was Galileo ostracised by the Catholic Church after he claimed that there were mountains on the Moon and that Jupiter had four moons of its own?

....

....

....

.... **[3 marks]**

3 Until the telescope was invented in the 17th century, astronomers observed the night sky with the naked eye.

Describe **two** advantages of using a telescope over the naked eye.

....

.... **[2 marks]**

4 Describe the nature and position of the asteroid belt.

....

.... **[2 marks]**

5 Why can the Hubble space telescope take much clearer photographs than any telescopes on Earth?

....

....

.... **[2 marks]**

Topic 1: 1.1, 1.2, 1.3, 1.4

Reflection, refraction and lenses

1 Use words from the list to complete the law of reflection.

| ray | normal | incidence | reflected | angle | mirror |

The angle of .... is equal to the ....

of reflection. The incident ray, the normal and the .... ray are all

in the same plane. The .... is an imaginary line at right angles to

the surface of the mirror where the reflection happens. **[4 marks]**

G–E

2 The diagram opposite shows a periscope.
Complete the diagram by drawing
in the path of the rays of light
from the tree towards the eye.

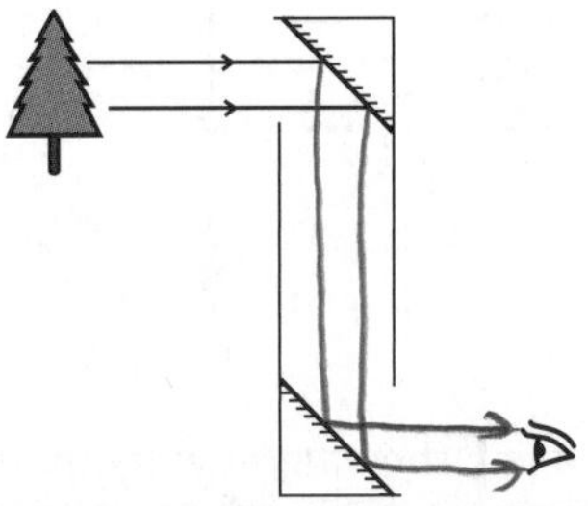

[3 marks]

3 Which of the following diagrams correctly shows how a ray of light will travel through a
rectangular glass block? Put a tick in the box below the correct diagram.

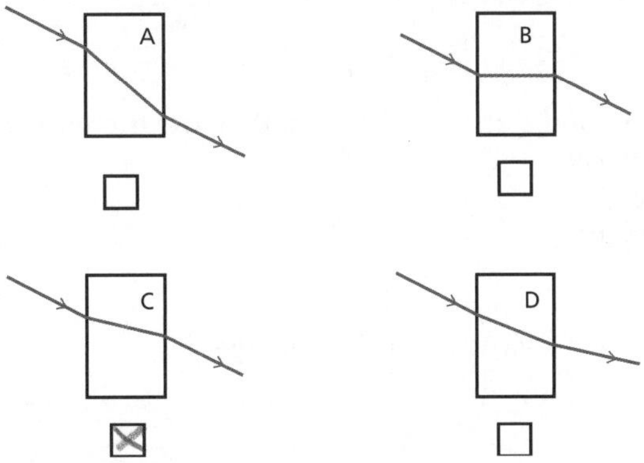

A

☐

B

☐

D–C

C

☒

D

☐

[1 mark]

4 The speed of light in different materials is shown in the table below. Which material refracts
light by the largest angle? Explain why.

Material	Speed of light (m/s)
Air	3.0×10^8
Crown glass	2.0×10^8
Water	2.3×10^8
Quartz	2.1×10^8

.... Crown glass ....

B–A*

.... **[2 marks]**

5 Converging lenses are used in many optical instruments, including microscopes and telescopes.

a Explain what is meant by the term 'focal point' of a converging lens.

....

.... **[2 marks]**

G–E

b Which of these two lenses will have the longer focal length?

A

B

.... **[1 mark]**

Lenses in telescopes

1 Describe a simple way to find the focal length of a converging lens.

....

....

.... [3 marks]

2 Look at the incomplete ray diagram below.

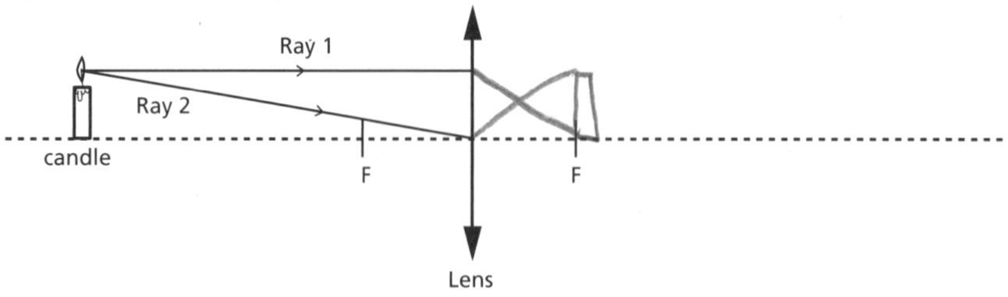

D–C

a Use a ruler to draw a line showing how the path of ray 1 is refracted through the lens. [2 marks]

b Use a ruler to draw a line showing the path of ray 2 as it passes through the lens. [2 marks]

c Mark the position of the image. [1 mark]

d Is the image magnified or diminished?

.... magnified .... [1 mark]

e Measure the object and the candle on your diagram, then use the formula below to calculate the magnification of the image.

$$\text{Magnification} = \frac{\text{image height}}{\text{object height}}$$

B–A*

.... [1 mark]

f Explain the difference between a virtual and a real image.

.... virtual is a mirror real is an existant image ....

.... [2 marks]

3 The diagram below shows a simple refracting telescope. Label the diagram with the following words:

G–E

eyepiece

real image

objective lens

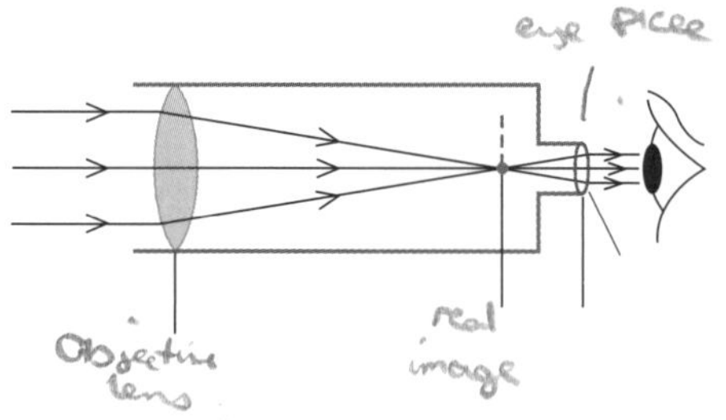

[3 marks]

4 Describe how a refracting telescope works.

D–C

....

....

.... [3 marks]

5 The Hubble space telescope is orbiting the Earth.

Is it a refracting or reflecting telescope? Give **two** reasons why this type of telescope was chosen for Hubble.

B–A*

....

....

.... [3 marks]

Waves

1 Complete the sentences describing waves, using words from the list below.

> vibrate   oscillations   energy   speed   matter

Waves transfer ...., without transferring

.... Waves are caused by .... [3 marks]

2 The diagram opposite shows a transverse wave.

Write the letter of the line that represents:

a The amplitude of the wave .... [1 mark]

b The wavelength of the wave .... [1 mark]

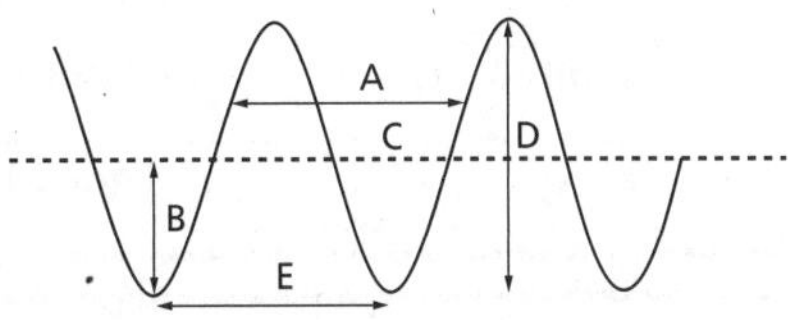

3 Tom is watching waves on a beach. He counts 15 waves in a minute. The distance between the crests is 8 m. Calculate the speed of the wave.

....

.... [3 marks]

4 Class 10C are trying to measure the speed of sound. Lizzie stands at the far side of the playing field, 200 m away, with a starting pistol. The rest of the class all record the time between seeing the smoke and hearing the shot.

a Explain why they all get slightly different results.

....

.... [2 marks]

b Here are five pupils' results: 0.55 s; 0.62 s; 0.58 s; 0.56 s; 0.58 s. Work out the average time, then calculate the speed of sound.

....

.... [3 marks]

5 Sound waves travel at 1500 m/s in water. If the frequency of a sound is 500 Hz, calculate the wavelength in water.

.... [2 marks]

6 There are two main types of wave: longitudinal and transverse. Draw lines joining the types of wave to the correct descriptions. [6 marks]

Longitudinal		Transverse

Sound wave

Vibrations at right angles to the direction of travel

Transfer energy

Light waves

Radio wave

7 There are two main types of seismic wave.

a What is a seismic wave?

....

.... [2 marks]

b Describe how a seismometer works.

....

.... [2 marks]

The electromagnetic spectrum

1 The electromagnetic spectrum is a family of waves.

a Complete the following statements about electromagnetic waves.

Electromagnetic waves are t.. waves. They all travel

through a v.. at the same speed. **[2 marks]**

b Put the following waves into the correct spaces in the electromagnetic spectrum:

infrared   **gamma rays**   **ultraviolet**   **radio waves**

	X-rays		Visible light		Microwaves	

[4 marks]

2 a Which type of wave has the longest wavelength?

.... **[1 mark]**

b Which type of wave has the most energy?

.... **[1 mark]**

c How do we protect ourselves from the harmful effects of ultraviolet waves?

....

.... **[2 marks]**

3 Explain what is meant by a spectrum of white light and describe how you can produce one in a school laboratory.

....

....

.... **[2 marks]**

4 a The physicist William Herschel carried out experiments to find out which colour of the spectrum was the warmest.

Explain how this led to the discovery of infrared waves.

....

....

.... **[3 marks]**

b Silver chloride turns black when it is exposed to sunlight. Johann Ritter carried out experiments to find out which colour of light reacted fastest with silver chloride.

Explain how this led to the discovery of ultraviolet waves.

....

....

.... **[3 marks]**

5 Electromagnetic waves travel at a speed of 3×10^8 m/s in a vacuum. If the frequency of an EM wave is 60 MHz:

a Calculate the wavelength of the wave.

....

.... **[2 marks]**

b Which type of electromagnetic wave is it?

.... **[1 mark]**

Uses of EM waves

1 Draw lines to join the type of electromagnetic radiation with its use.

| Microwaves | | Mobile-phone communication |

| X-rays | | TV remote control |

G–E

| Infrared | | Airport security scanner |

B–A*

| Ultraviolet |

| Water sterilisation |

[4 marks]

2 You phone your friend in the USA on your mobile phone. The sound is coded and sent using microwaves.

Explain how the microwave signal travels to the USA.

....

....

.... **[3 marks]**

3 All objects emit some infrared waves.

Explain how infrared imaging can be used to compare home insulation.

Explain how you can tell that one house has much better wall insulation than the others.

....

....

.... **[3 marks]**

4 a On each diagram below, draw the path of the ray of light as it leaves the semicircular block.

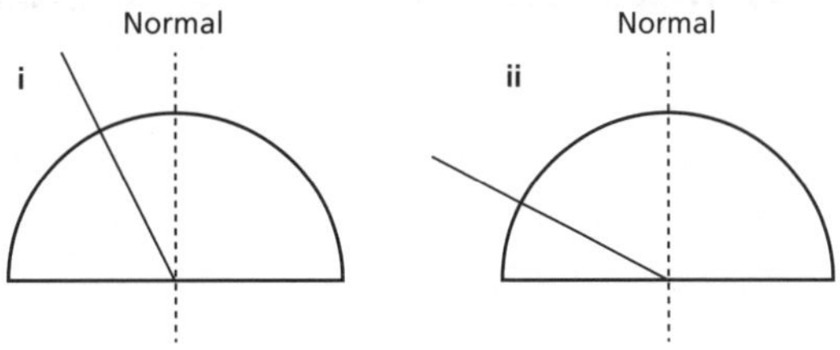

Normal   Normal

i   ii

[4 marks]

b The diagram shows light travelling down an optical fibre.

Continue the path of the light ray until it emerges from the optical fibre.

[3 marks]

Gamma rays, X-rays, ionising radiation

1 X-rays are dangerous, but they are used in hospitals.

Explain why they are used, despite being harmful.

G–E

....

.... **[2 marks]**

2 Gamma rays and X-rays have similar wavelengths and frequencies.

Name **one** similarity and **one** difference between gamma rays and X-rays.

D–C

....

.... **[2 marks]**

3 Describe how X-rays are produced.

B–A*

....

....

.... **[3 marks]**

4 To a physicist, ionising radiation can mean electromagnetic waves or radiation from the nucleus of atoms.

Name the **three** different types of ionising radiation that come from the nucleus of atoms.

G–E

.... **[3 marks]**

5 Joanna and Michael are investigating the ionising radiation emitted from some rocks. Joanna suggests that they lower the temperature of the rocks because this will reduce the amount of radiation emitted. Michael disagrees with her, and says that lowering the temperature will make no difference.

a Who is correct? Explain your answer.

D–C

....

....

.... **[3 marks]**

b Joanna and Michael use a Geiger–Müller tube to measure the radioactivity emitted from the rocks.

Describe how a Geiger–Müller tube detects ionising radiation.

B–A*

....

....

....

.... **[4 marks]**

Topic 2: 2.5, 2.6a, b, d, 2.7f, g, 2.8, 2.9

The Universe

1 List the objects below in order of size, from smallest to largest:

Jupiter   Mercury   Moon   Sun   comet   Milky Way

G–E

....

.... [3 marks]

2 The table below contains information about some of the bodies in the Solar System.

Use the data to answer the questions below.

Body	Average distance from Sun (AU)	Diameter relative to Earth
Mercury	0.39	0.38
Earth	1.00	1.00
Mars	1.52	0.53
Jupiter	5.20	11.20
Moon	1.00	0.27

D–C

a How many times larger than Mercury is Jupiter?

.... [1 mark]

b How many times further away from the Sun than Mars is Jupiter?

.... [1 mark]

c How many times larger than the Moon is the Earth?

.... [1 mark]

3 Light from the Sun takes approximately 8 minutes to get to Earth. The speed of light is 3×10^8 m/s.

a Calculate the average distance of the Earth from the Sun. Show your working.

....

....

B–A*

.... [3 marks]

b Explain why looking at distant stars is like looking back in time.

....

.... [2 marks]

4 Scientists have studied space for thousands of years.

Early telescopes only used light to obtain images of space. Why do we now use other parts of the electromagnetic spectrum, such as radio waves, to study space?

G–E

....

....

.... [2 marks]

5 a Scientists have sent a robotic lander to Mars.

What sort of information can a lander find out about Mars?

....

.... [2 marks]

b What is the Search for Extraterrestrial Intelligence (SETI) project?

D–C

....

....

....

.... [3 marks]

Analysing light

1 a What is a spectrometer?

.... [1 mark]

b Describe how to make a simple spectrometer.

....

....

....

.... [3 marks]

G–E

2 Astronomers use spectrometers to analyse the light from stars.

a Why do different stars show different spectra?

....

.... [2 marks]

b What information can astronomers obtain by studying the spectra from stars?

....

.... [2 marks]

D–C

3 Earth's atmosphere protects us from harmful radiation from space.

Use the graph opposite to explain why holes in the ozone layer have been linked to a rise in the number of cases of skin cancer.

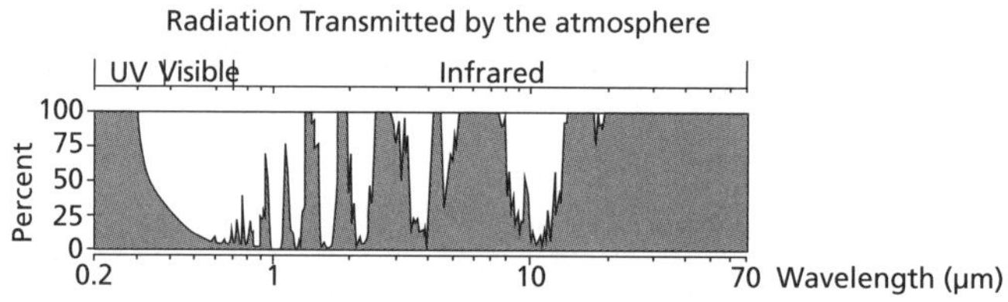

Radiation Transmitted by the atmosphere

....

....

.... [3 marks]

B–A*

4 The left-hand diagram opposite shows some circular wave fronts from a stationary source.

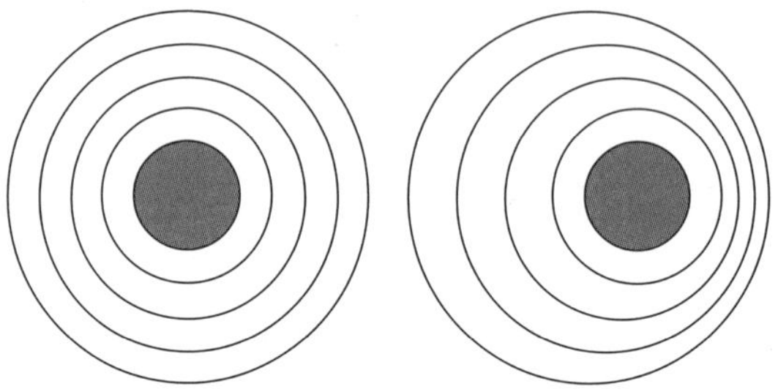

A

a What is the wavelength of the waves in mm?

.... [1 mark]

b The right-hand diagram shows the same wave fronts, but now the source is moving to the right. If you were observing the waves from position A, the wave source would be moving towards you.

What has happened to the wavelength?

....

.... [1 mark]

c If the waves were sound waves, what would happen to the pitch of the sound?

.... [1 mark]

D–C

5 Explain what is meant by red-shift.

....

....

.... [3 marks]

B–A*

The life of stars

1 Put the following stages in the birth of a star into the correct order.

The first one has been done for you.

A A nebula contains clouds of dust, ice and gas.

B All the clouds disappear and a glowing ball of gas remains.

C The particles start to collide with each other and the temperature starts to rise.

D Nuclear fusion reactions occur, giving out energy in the form of electromagnetic radiation.

E The force of gravity attracts all the matter together

A .... **[3 marks]**

G–E

---

2 Our Sun is a main sequence star in a steady state.

  a Explain why the Sun does not collapse in on itself due to the force of gravity.

  ....

  .... **[2 marks]**

  b Towards the end of its life, the Sun will cool and expand to become a red giant.

    What will happen once the fuel for the nuclear reactions runs out?

  ....

  .... **[2 marks]**

D–C

---

3 Massive stars like Rigel in the constellation Orion look blue in the night sky.

  a Why do they look bluer than average-sized stars like the Sun?

  .... **[1 mark]**

  b Describe what will happen to Rigel when it comes to the end of its life.

  ....

  ....

  ....

  .... **[4 marks]**

B–A*

---

4 Put these objects in order of size, from smallest to largest:

      supernova   **white dwarf**   **red giant**   **main sequence star**

  ....

  .... **[3 marks]**

D–C

---

5 Scientists have predicted that there is a black hole in the centre of the Milky Way.

Explain why it is difficult to observe a black hole.

  ....

  ....

  .... **[2 marks]**

B–A*

Theories of the Universe

1 In the 1940s there were two opposing theories of the Universe: the Steady State theory and the Big Bang theory.

Draw lines to join the main theories to the correct ideas.

Big Bang theory

Steady State theory

| 1 The Universe spontaneously creates matter in empty space. |
| 2 The Universe is expanding. |
| 3 The Universe is constantly changing. |
| 4 The Universe had a beginning about 14 billion years ago, and will eventually end. |
| 5 The Universe has no beginning and no end. |

[6 marks]

2 a Describe the evidence that proves that the Universe is expanding.

[2 marks]

b What is meant by 'cosmic background radiation'?

[2 marks]

c Why does the observation of cosmic background radiation support the Big Bang theory?

[2 marks]

3 Read the paragraph below about the discovery of cosmic background radiation.

In 1964, Penzias and Wilson accidentally discovered the existence of cosmic background radiation. They were actually measuring radio waves reflected from orbiting satellites with a very large horn antenna. They found some 'interference', which they could not account for. They eliminated all known sources of the radiation, but still recorded microwave interference. They believed it might be caused by bird droppings inside the antenna, so they spent several hours cleaning it off. They still recorded the 'interference', and found that it was equal in strength in all directions. They eventually linked this to Gamov's predicted cosmic background radiation.

a What is meant by 'interference'?

[2 marks]

b Why did Penzias and Wilson need to eliminate all known sources of microwave radiation?

[2 marks]

c Why is cosmic background radiation equal in strength in all directions?

[2 marks]

d Why should cosmic background radiation appear in the microwave section of the electromagnetic spectrum?

[3 marks]

Topic 3: 3.14, 3.15, 3.16, 3.19, 3.20, 3.21, 3.22

Ultrasound and infrasound

1 Sounds can have a range of frequencies.

a Which of these frequencies can most humans hear: 6 Hz, 50 Hz, 2000 Hz, 50000 Hz?

.... [2 marks]

b As you get older, your hearing deteriorates. Which frequencies do you lose the ability to hear – higher or lower?

.... [1 mark]

c Explain what is meant by the terms infrasound and ultrasound.

....

.... [2 marks]

G–E

2 Ultrasound is used to scan unborn babies in the womb.

a Describe how ultrasound can be used to create an image.

....

.... [3 marks]

b Why is ultrasound used to scan unborn babies?

....

.... [2 marks]

3 Name **one** man-made and **one** natural source of infrasound.

....

.... [2 marks]

D–C

4 A boat sends out pulses of ultrasonic waves and receives echoes. Both the outgoing pulses and reflected pulses are displayed on an oscilloscope screen as shown below.

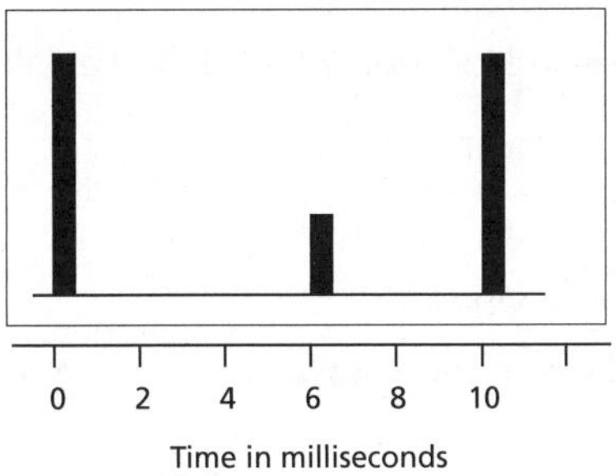

Time in milliseconds

a How long did it take the sound to travel to the sea bed and back again?

.... [1 mark]

b If the speed of sound in water is 1500 m/s, calculate the depth of water below the bottom of the boat.

....

.... [3 marks]

c How will the pattern on the screen change as the tide comes in?

.... [2 marks]

d Suggest how you could adjust the equipment so that greater depths could be measured.

.... [1 mark]

B–A*

Earthquakes and seismic waves

1 The Earth is made up of several layers.

   a Label the diagram showing four layers of the Earth.

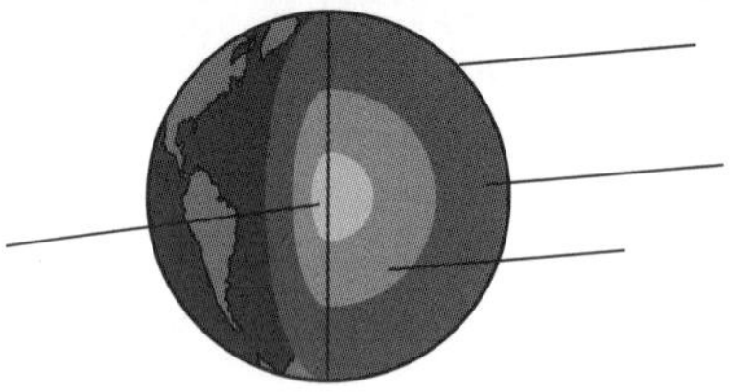

        [4 marks]

   b What is magma?

   ....

        [1 mark]

2 The Earth's crust is made up of several solid plates that float on top of the magma.

   a What is the name of these plates?

   .... [1 mark]

   b How do the plates cause earthquakes?

   ....

   ....

   .... [3 marks]

   c Alfred Wegener first suggested this theory about Earth's plates in 1915.

      Describe the evidence that supports his theory.

   ....

   ....

   ....

   .... [3 marks]

3 There are two main types of seismic wave: S waves and P waves.

   a Which seismic waves are longitudinal? .... [1 mark]

   b Which seismic waves can travel through Earth's liquid outer core? .... [1 mark]

   c Which seismic waves can travel through Earth's semi-solid mantle?................... [1 mark]

4 It is difficult to predict the exact time and location an earthquake will occur. Explain why.

   ....

   .... [2 marks]

5 The trace from a seismograph is shown opposite.

   a Which vibration is the P wave and which the S wave?

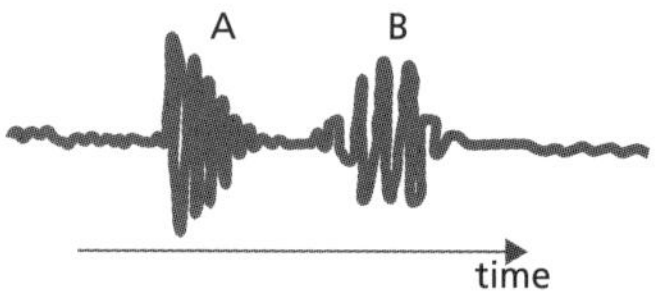

   .... [1 mark]

6 The diagram opposite shows how P waves travel through the centre of the Earth.

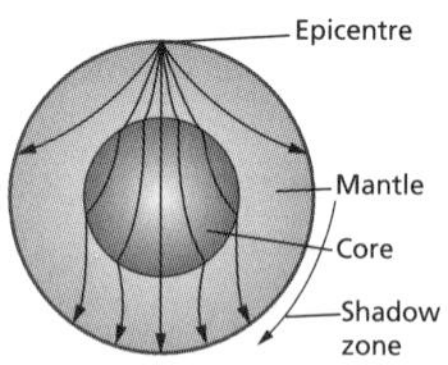

   a Explain why there is a shadow zone where no P waves are detected.

   ....

   .... [2 marks]

   b Complete the diagram opposite to show how S waves travel through the centre of the Earth.

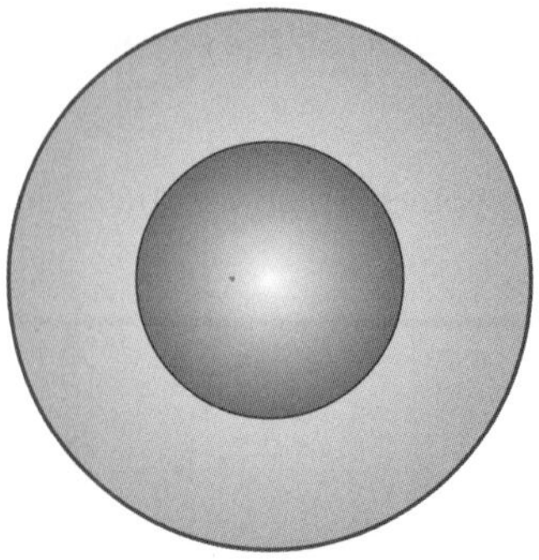

        [3 marks]

Electrical circuits

1 a Draw an electric circuit for a bulb with a battery and a switch.

G-E

[3 marks]

b Describe what happens to the charged particles in the above circuit.

....

....

....

D-C

[3 marks]

2 In the circuit opposite, both the bulbs are identical.

What is the current at positions X, Y and Z?

0.6A   Z

X

Y

B-A*

X = ...

Y = ...

Z = ...

[3 marks]

3 In the circuit below, bulb A has a voltage of 3 V across it, and bulb B has a voltage of 4 V across it.

A   B

G-E

Which bulb, A or B, is converting the most energy?

.... [1 mark]

4 Complete the following sentences using the words **current** or **voltage**.

When lamps are connected in series, the .... through them is

the same. The .... across each lamp will add up to the total

supply ....

When lamps are connected in parallel, the .... across each

bulb will be the same as the .... across the power supply.

D-C

[3 marks]

5 What is meant by the term 'potential difference'?

....

....

....

B-A*

[3 marks]

Electrical power

1 The power ratings of some domestic appliances are listed in the table below.

Appliance	Power rating in watts	Average daily use
Kettle	2000	10 minutes
Iron	1200	30 minutes
Vacuum cleaner	1000	20 minutes

a Which appliance uses the most energy per second?

.... [1 mark]

b Which appliance uses the most energy in an average day?

.... [1 mark]

c Which appliance will cost the most to run?

.... [1 mark]

d Complete the table below showing the amount of energy used by each appliance in both joules and kilowatt-hours.

Appliance	Energy used in joules	Energy used in kW h
Kettle		
Iron		
Vacuum cleaner		

[6 marks]

e Explain why electricity supply companies use kW h instead of joules as units of energy.

.... [1 mark]

f The mains electricity supply is 230 V. Calculate the current that each appliance will use, and decide which size fuse should be used for each appliance. Fuses available are 1 A, 3 A, 5 A and 13 A.

Kettle .... [2 marks]

Iron .... [2 marks]

Vacuum cleaner .... [2 marks]

2 The local council is encouraging people to save energy in the home. One suggestion is to use energy-saving light bulbs instead of filament bulbs.

a Give **two** advantages of using energy-saving light bulbs.

....

.... [2 marks]

b Mr Hicks says that he does not want to use energy-saving light bulbs as they cost about twice as much as filament bulbs.

Explain the idea of 'payback time', to encourage him to use them.

....

....

.... [3 marks]

3 The table below shows the initial cost of installing some other energy-saving devices in the home.

a Complete the table below to show the annual savings and payback time for each device.

Device	Initial cost (£)	Annual saving (£)	Payback time (years)
Double glazing	7000	350	
Loft insulation	450		6
Draught excluders	40	5	
Cavity-wall insulation	550		5

[4 marks]

b Which device has the longest payback time? Give **one** other reason why householders should choose to install this device.

....

.... [2 marks]

Energy resources

1 Fossil fuels are used to produce electricity in power stations. Put the following sentences in the correct order to describe how they are used. The first one has been done for you.

A The fuel is burnt to produce heat.   **D** The steam turns turbines.

B The turbines are connected to generators.   **E** The heat is used to boil water to produce steam.

C The generators produce electricity.

A, ... , ... , ... , ...   **[3 marks]**

(G–E)

2 **a** Circle the two gases that are produced when fossil fuels are burnt.

               oxygen   **carbon dioxide**   **ammonia**   **sulfur dioxide**   **[2 marks]**

  b What problems are associated with these two gases?

      ....

      .... **[2 marks]**

(D–C)

  c Nuclear power stations do not produce these gases. Name **two** disadvantages of using nuclear power stations.

      ....

      .... **[2 marks]**

3 Explain what is meant by the term 'greenhouse effect'.

   ....

   ....

   .... **[3 marks]**

(B–A)*

4 Next to each of the following energy resources, write whether it is renewable or non-renewable.

wind	....	**biomass**	....
coal	....	**nuclear**	....
hydroelectric	....	**oil**	....
solar	....	**tidal**	.... **[4 marks]**

(G–E)

5 **a** Explain briefly how wind turbines are used to create electricity.

      ....

      .... **[2 marks]**

  b Name **two** disadvantages of wind turbines.

      ....

      .... **[2 marks]**

(D–C)

6 The graph opposite shows how the power output from a wind turbine varies with wind speed.

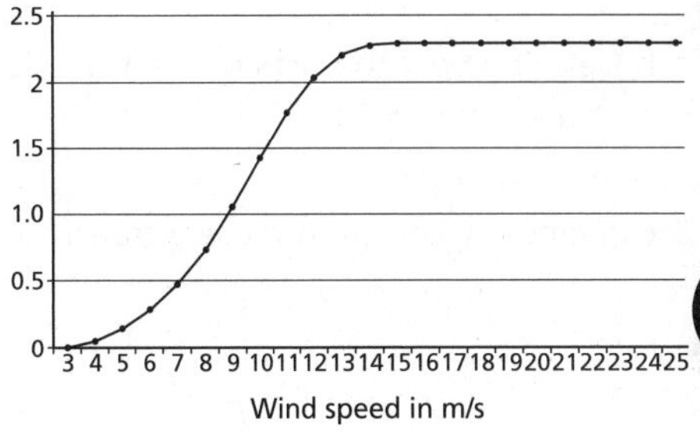

  a What is the maximum power output of the wind turbine?

      .... **[1 mark]**

  b At what wind speed is the maximum power obtained?

      .... **[1 mark]**

Wind speed in m/s

  c The graph only shows data from wind speeds of about 4 m/s up to 25 m/s. Suggest why only this range of data has been given.

      ....

      .... **[2 marks]**

(B–A)*

Generating and transmitting electricity

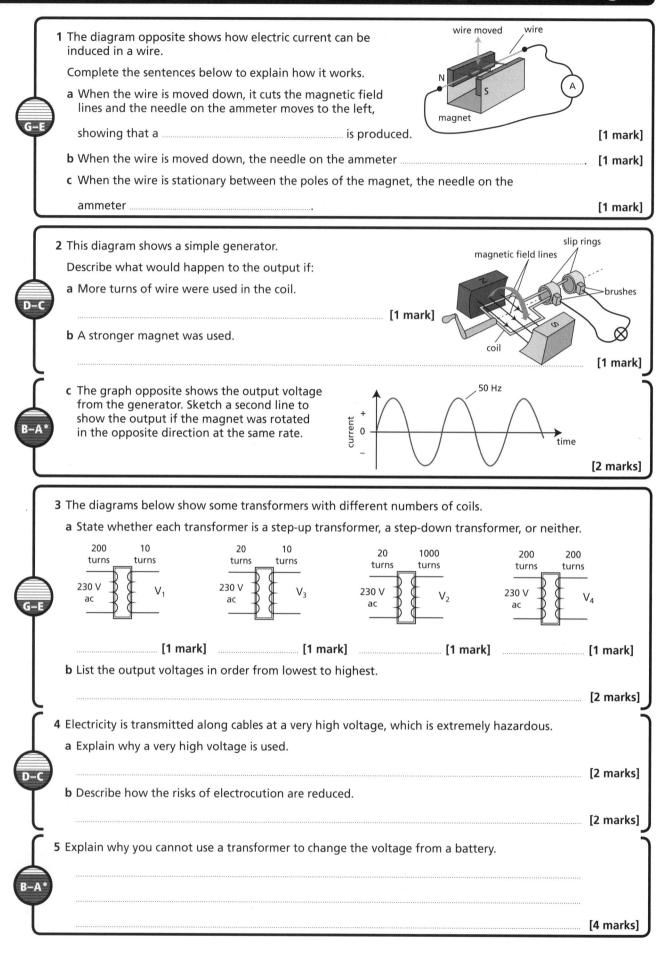

1 The diagram opposite shows how electric current can be induced in a wire.

Complete the sentences below to explain how it works.

a When the wire is moved down, it cuts the magnetic field lines and the needle on the ammeter moves to the left,

showing that a .... is produced. **[1 mark]**

b When the wire is moved down, the needle on the ammeter .... **[1 mark]**

c When the wire is stationary between the poles of the magnet, the needle on the

ammeter .... **[1 mark]**

2 This diagram shows a simple generator.

Describe what would happen to the output if:

a More turns of wire were used in the coil.

.... **[1 mark]**

b A stronger magnet was used.

.... **[1 mark]**

c The graph opposite shows the output voltage from the generator. Sketch a second line to show the output if the magnet was rotated in the opposite direction at the same rate.

[2 marks]

3 The diagrams below show some transformers with different numbers of coils.

a State whether each transformer is a step-up transformer, a step-down transformer, or neither.

| 200 turns / 10 turns | 20 turns / 10 turns | 20 turns / 1000 turns | 200 turns / 200 turns |
| 230 V ac V_1 | 230 V ac V_3 | 230 V ac V_2 | 230 V ac V_4 |

.... **[1 mark]** .... **[1 mark]** .... **[1 mark]** .... **[1 mark]**

b List the output voltages in order from lowest to highest.

.... **[2 marks]**

4 Electricity is transmitted along cables at a very high voltage, which is extremely hazardous.

a Explain why a very high voltage is used.

.... **[2 marks]**

b Describe how the risks of electrocution are reduced.

.... **[2 marks]**

5 Explain why you cannot use a transformer to change the voltage from a battery.

....
....
....

[4 marks]

Energy and efficiency

1 Draw a line to match the object to the type of energy.

A roller coaster at the top of the ride	Elastic potential energy
A stretched rubber band	Gravitational potential energy
A battery	Heat energy
A cup of tea	Chemical energy

[4 marks]

2 Complete the table below, identifying the energy input, useful energy output and the wasted energy output for a variety of devices. Some have been done for you.

Device	Energy input	Useful energy output	Wasted energy output
Electric fan	Electricity		
Television		Light and sound	
Catapult		Kinetic	Heat
Gas ring on a cooker			Light and sound

[7 marks]

3 Explain what is meant by the principle of conservation of energy.

....

.... [3 marks]

4 The diagram below shows how a coal-fired power station generates electricity.

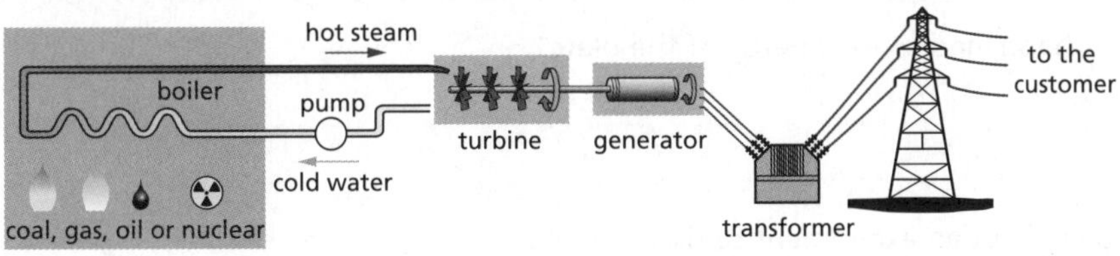

a Complete the energy transfer chain for this process.

.... energy in the coal

→

.... energy in the steam

.... energy in the turbine

→

.... energy from the generator

[4 marks]

b For every 360 MJ of energy stored in the coal, only 144 MJ of electricity is generated. Calculate the efficiency of the power station.

.... [2 marks]

c What happens to the rest of the energy from the coal?

.... [2 marks]

5 Sketch a labelled Sankey diagram for the power station in Question 4.

[4 marks]

Radiated and absorbed energy

1 a Which of the following contains the most heat energy?

   i 100 ml of water at 20 °C   **ii** 200 ml of water at 40 °C

   iii 100 ml of water at 40 °C

   .... [1 mark]

b Which of the following would emit the most heat energy?

   i A cup of tea in a black cup   **ii** A cup of tea in a white cup

   iii A cup of milky tea in a white cup

   .... [1 mark]

c Which of the following would absorb the most heat energy on a sunny day?

   i A black water bottle   **ii** A transparent water bottle

   iii A white water bottle

   .... [1 mark]

2 Ellie got an ice cube out of the freezer (at about –5 °C) and put it on a plate (at about 50 °C), which she had just got out of the dishwasher. The room temperature in her kitchen was 20 °C. She left the ice cube there for several hours and the ice melted to a pool of water.

a What was the final temperature of the plate? .... [1 mark]

b What was the final temperature of the water? .... [1 mark]

c Describe what happened to the energy of the plate?

....

.... [3 marks]

3 Sam and Joe carried out an experiment to find out what type of surface lost the most heat energy. They coated one beaker with silver paint and a second beaker with matt black paint.

They put some hot water into each beaker and recorded the temperature at intervals as the water cooled down. The results for the silvered beaker were as follows:

Time (minutes)	Temperature of water (°C)
0	80
5	71
10	65
15	61
20	58
25	56
30	55

a Plot a line of time on the x-axis against temperature on the y-axis on the grid provided.   [4 marks]

b Name **two** ways in which Sam and Joe could make sure the test was fair.

....

.... [2 marks]

c Sketch a line on the graph showing the results for the blackened beaker.   [2 marks]

d Explain your answer to part C.

....

.... [2 marks]

P1 Extended response question

*Our Sun was born in a nebula almost five billion years ago. Our Sun is an average-sized star and it will eventually die.

Describe the main stages in an average-sized star's life.

[6 marks]

Questions labelled with **asterisk** (*) are ones where the quality of your written communication will be assessed. You should take particular care with your spelling, punctuation and grammar, as well as the clarity of expression in these answers.

Electrostatics

1 Scientists believe that all matter is made of atoms.

  a Which two particles are present in the nucleus of an atom?

    .... **[2 marks]**

  b Explain why the nucleus has a positive charge.

    ....

    .... **[2 marks]**

2 When Brendan takes off his jumper it crackles. Electrostatic charge has built up on his jumper.

  a Explain how this could have happened.

    .... **[1 mark]**

  b Which particles cause this crackling sound?

    .... **[1 mark]**

  c After Brendan has removed his jumper, he finds that his hair has become charged.
Explain why the strands of hair repel each other.

    ....

    .... **[2 marks]**

3 If you rub a balloon against your clothes and then hold it against a wall, it sticks.

  a Explain how the transfer of charge causes this to happen.

    ....

    ....

    .... **[3 marks]**

  b What name is given to the separation of charges by another charged object?

    .... **[1 mark]**

G–E

4 An atom can be charged or uncharged.

  a What can you say about the number of protons and electrons in an uncharged atom?

    .... **[1 mark]**

  b What name is given to a charged atom?

    .... **[1 mark]**

  c Explain how an atom becomes positively charged.

    .... **[1 mark]**

D–C

5 When a charged glass rod is brought near to the cap of a gold-leaf electroscope, the leaf moves away from the metal rod. Explain why this occurs.

    ....

    ....

    ....

    .... **[3 marks]**

B–A*

Uses and dangers of electrostatics

1 A farmer uses an electrostatic plant sprayer to spray his crops with insecticide.

Give **three** advantages to using an electrostatic sprayer.

D–C

[3 marks]

2 The diagram below shows a negatively charged sphere with an electric field around it.

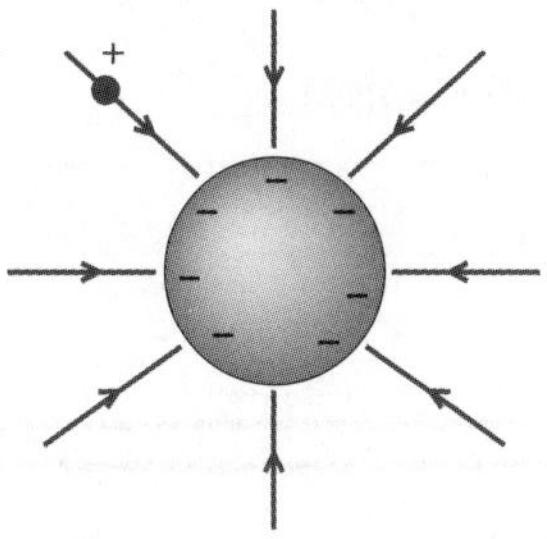

B–A*

a What name is given to the lines?

[1 mark]

b If the lines were closer together, what change would this represent?

[1 mark]

3 Adrian brings his finger close to a charged insulator.

Describe what Adrian will experience and why.

G–E

[2 marks]

4 A lorry is fuelled using a plastic fuel pipe.

a Explain how static electricity could cause an explosion during refuelling.

[2 marks]

D–C

b How could a metal cable be used to reduce the risk of such an explosion?

[2 marks]

5 A lightning conductor can help protect a tall building from lightning strikes.

a The top of the conductor points upwards from the top of the building. Where is the other end of the conductor?

[1 mark]

B–A*

b A negatively charged cloud passes overhead. What charge is induced at the top of the spike?

[1 mark]

Current, voltage and resistance

1 What is meant by 'direct current'?

.... **[1 mark]**

2 In the series circuit below, the current at X is 2.0 A.

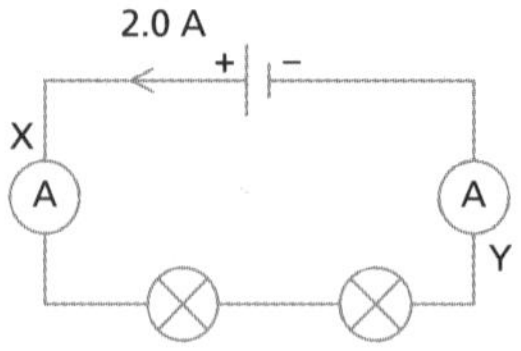

a What current passes through each of the lamps?

.... **[1 mark]**

b What is the value of the current at Y?

.... **[1 mark]**

3 The diagram opposite shows a circuit with a cell, a resistor and two lamps.

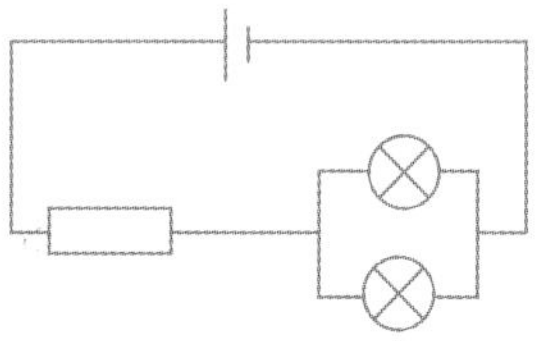

a Are the lamps connected in series or in parallel with one another?

.... **[1 mark]**

b Is the resistor connected in series or in parallel with the two lamps?

.... **[1 mark]**

4 A voltmeter can be used to measure the potential difference across a component.

a How should a voltmeter be connected in the circuit?

.... **[1 mark]**

b What second measuring device is needed in order to calculate the resistance of the component?

.... **[1 mark]**

c How should this second device be connected in the circuit?

.... **[1 mark]**

5 A torch lamp has a resistance of 3.5 Ω.

What is the potential difference across it when a current of 2.0 A flows through it?

.... **[1 mark]**

6 A resistor and a lamp are connected in series in a circuit.

a The potential difference across the resistor is 1.5 V. What energy is transferred to heat in the resistor when 10 C of charge flows through it?

.... **[1 mark]**

b The lamp transfers 3.0 J of energy into heat and light when the same charge passes through it. What is the potential difference across the lamp?

.... **[1 mark]**

Topics 1 and 2: 1.9, 1.10, 1.11, 1.12, 1.13, 2.1, 2.2, 2.3, 2.4, 2.5, 2.8

Lamps, resistors and diodes

1 Doris wants to investigate how the resistance of a wire depends on its length.

a What two quantities should Doris measure so that she can calculate the resistance of the wire?

.... [2 marks]

b Complete the circuit diagram below that Doris could use for her investigation.

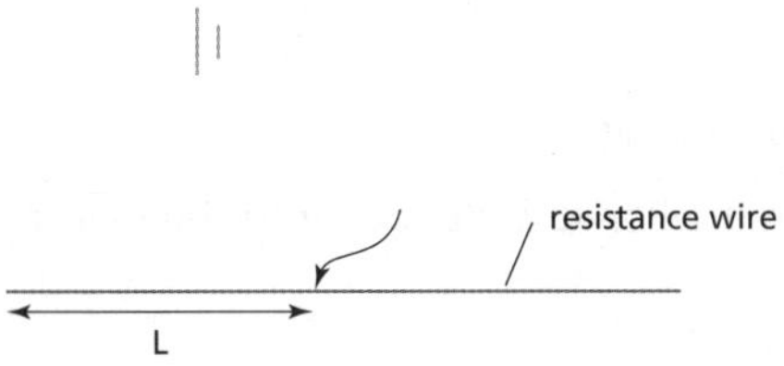

resistance wire
L

[3 marks]

G–E

c What relationship should Doris find that relates the resistance and the length of the wire?

....

....

.... [2 marks]

2 The graph of current against voltage for a component is a straight line.

What does this tell you about the resistance of the component?

.... [1 mark]

3 The resistance of a filament lamp increases as the current increases.

a Explain why this happens.

....

.... [2 marks]

D–C

b What would a graph of current against voltage for a filament lamp look like?

....

.... [2 marks]

4 Explain how the resistance of a diode changes with the current through it.

....

....

.... [3 marks]

B–A*

Heating effects, LDRs and thermistors

1 Tulsey purchases a projector to use with her laptop. The instructions say that when the projector is switched off after use, the power must not be disconnected from the projector until the fan has stopped.

Explain the reason for this.

....

....

....

[3 marks]

2 An electric heater has a power of 4400 W.

a What current will flow through it when it is plugged into the mains supply of 220 V? Show your working.

....

....

[3 marks]

b What energy will be transferred by the heater in 5 minutes? Show your working.

....

....

[3 marks]

3 When current flows through a wire, electrons travel through the lattice of the metal.

Explain how this causes the wire to heat up.

....

....

....

....

[4 marks]

4 In what ways does the resistance of a thermistor depend on temperature?

....

....

[2 marks]

5 Suggest whether it would be more suitable to use an LDR or a thermistor in the following circuits.

a A central-heating controller.

....

[1 mark]

b A circuit for automatically controlling street lamps.

....

[1 mark]

c A circuit to control the cooling fan in a laptop.

....

[1 mark]

6 In most metals, increasing the temperature increases the resistance of the metal. In a thermistor, increasing the temperature decreases the resistance.

Explain why this is so.

....

....

....

[2 marks]

Scalar and vector quantities

1 Explain the difference between a scalar quantity and a vector quantity.

....

.... [2 marks]

2 Reggie sets out from home on his bicycle. First he goes to the shop, which is 500 m away, then he rides to his friend's house, which is a further 1200 m. He then returns home.

a What total distance has Reggie travelled when he returns home?

.... [1 mark]

b What is his displacement when he is at his friend's house?

.... [1 mark]

c What is his displacement when he returns home?

.... [1 mark]

3 Jesse Rose walks along a footpath for 6 minutes. She then cuts across a field, which takes a further 4 minutes. She walks a total distance of 900 m.

What is her average speed?

....

.... [2 marks]

4 A racing car accelerates from rest to 30 m/s in 6.0 s.

a What is its average acceleration?

.... [1 mark]

b The car continues to accelerate at an acceleration of 3.0 m/s^2.

What is the car's speed after a further 2.0 s?

.... [1 mark]

c The car then turns a corner at a constant speed.

Explain why the car is still accelerating.

....

.... [2 marks]

5 A light aeroplane has a deceleration of 2.5 m/s^2, and it is able to hit the runway at a minimum speed of 35 m/s. The runway of a remote airport is 400 m long.

a How many seconds will the plane take to stop?

.... [1 mark]

b Will the plane be able to land without overrunning the runway? Explain your answer.

....

....

.... [2 marks]

G–E

D–C

B–A*

Distance–time and velocity–time graphs

1 a What does the gradient of a distance–time graph represent?

.... **[1 mark]**

b A graph of distance against time for a car is a curved line with a decreasing gradient.

What does this tell you about the motion of the car?

.... **[1 mark]**

2 A graph representing a ball rolling down a slope from rest has an increasing gradient. By taking a tangent to the line, Xenia finds that the gradient at 0.40 s is 1.6 m/s.

Find the acceleration of the ball. Show your working.

....

.... **[2 marks]**

3 On a velocity–time graph, what does a horizontal line represent?

.... **[1 mark]**

4 The graph below shows how the velocity of a motorbike varies with time.

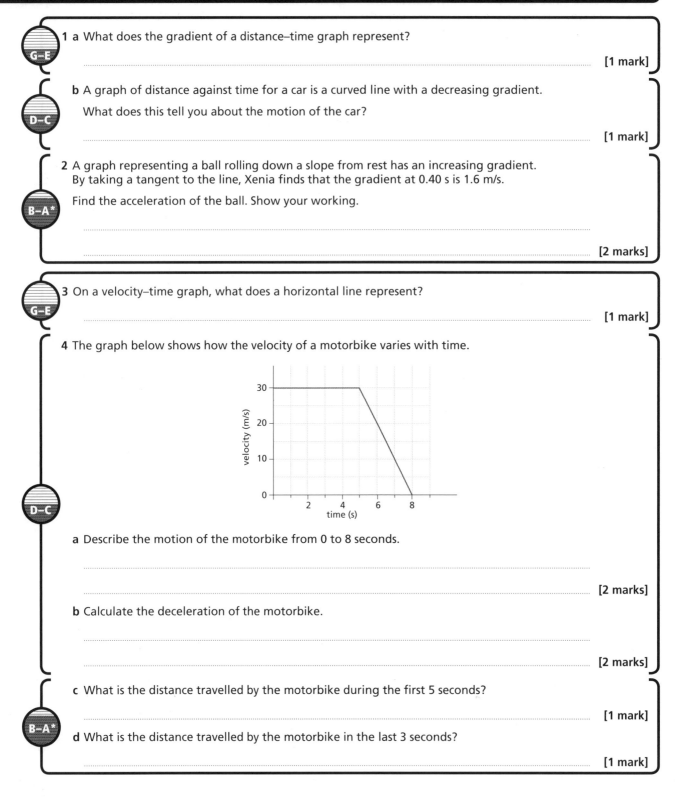

a Describe the motion of the motorbike from 0 to 8 seconds.

....

.... **[2 marks]**

b Calculate the deceleration of the motorbike.

....

.... **[2 marks]**

c What is the distance travelled by the motorbike during the first 5 seconds?

.... **[1 mark]**

d What is the distance travelled by the motorbike in the last 3 seconds?

.... **[1 mark]**

Understanding forces

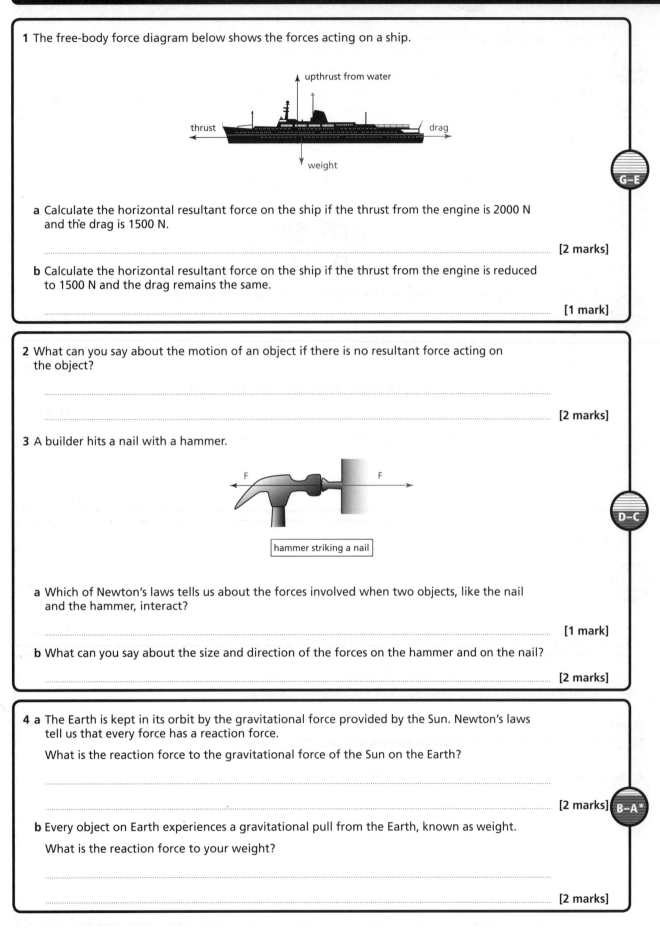

1 The free-body force diagram below shows the forces acting on a ship.

upthrust from water

thrust

drag

weight

G–E

a Calculate the horizontal resultant force on the ship if the thrust from the engine is 2000 N and the drag is 1500 N.

.... **[2 marks]**

b Calculate the horizontal resultant force on the ship if the thrust from the engine is reduced to 1500 N and the drag remains the same.

.... **[1 mark]**

2 What can you say about the motion of an object if there is no resultant force acting on the object?

....

.... **[2 marks]**

3 A builder hits a nail with a hammer.

F

F

hammer striking a nail

D–C

a Which of Newton's laws tells us about the forces involved when two objects, like the nail and the hammer, interact?

.... **[1 mark]**

b What can you say about the size and direction of the forces on the hammer and on the nail?

.... **[2 marks]**

4 a The Earth is kept in its orbit by the gravitational force provided by the Sun. Newton's laws tell us that every force has a reaction force.

What is the reaction force to the gravitational force of the Sun on the Earth?

....

.... **[2 marks]** B–A*

b Every object on Earth experiences a gravitational pull from the Earth, known as weight.

What is the reaction force to your weight?

....

.... **[2 marks]**

Force, mass and acceleration

1 Marylena wants to buy a new speedboat to participate in a speedboat race next month. She is interested in how fast the speedboat is able to accelerate.

What **two** factors will affect the acceleration of the boat?

.... [2 marks]

2 Cassandra wants to investigate the link between force and acceleration. She will use different elastic bands to vary the force on a trolley.

elastic bands

masses

trolley

a What other important piece of equipment is necessary for the investigation?

.... [1 mark]

b What quantity must she keep constant during the experiment?

.... [1 mark]

c What relationship between force and acceleration should she expect to find?

.... [1 mark]

3 A cyclist on his bike has a mass of 90 kg. The cyclist exerts an average force of 120 N as he pedals and he experiences a drag force of 30 N.

What is his average acceleration?

.... [2 marks]

4 What is the weight on Earth of:

a A horse of mass 80 kg?

.... [1 mark]

b A small dog of mass 6 kg?

.... [1 mark]

5 A skydiver is falling at a steady speed. He opens his parachute.

a What force dramatically increases when he opens the parachute?

.... [1 mark]

b In what direction is the resultant force on the skydiver at this point?

.... [1 mark]

c How does this affect his speed?

.... [1 mark]

d Explain how he reaches a new steady speed.

....

....

.... [3 marks]

Stopping distance

1 A car is travelling at 20 m/s. The driver sees a deer on the road and hits the brakes.

a What is the name given to the time between the moment that she sees the deer and the moment that she hits the brakes?

.... [1 mark]

b Find the distance travelled during this time if it takes the driver 0.3 s to apply the brakes.

.... [1 mark]

2 Complete the following equation:

Stopping distance = ... + ... [2 marks]

3 Give **three** factors that could increase the braking distance of a car.

....

....

.... [3 marks]

4 Friction opposes motion.

a Name a situation in which friction is useful.

.... [1 mark]

b Name a situation when friction is a nuisance and explain why.

.... [2 marks]

5 The graph below shows the thinking and braking distances for a car with an initial velocity of 20 m/s.

a Find the thinking distance from the graph.

.... [1 mark]

b Find the braking distance from the graph.

.... [1 mark]

c If the same car was travelling at 30 m/s, what would the thinking distance be?

.... [1 mark]

d If the same car was travelling at 30 m/s, what would the braking distance be? (Assume that all other conditions are the same.)

.... [1 mark]

e How much greater would the total stopping distance be for the car travelling at 30 m/s compared to 20 m/s?

.... [1 mark]

Momentum

1 A 200-kg truck is travelling at 12 m/s.

a What is its momentum?

.... [1 mark]

b Momentum is a vector quantity. What does this mean?

.... [1 mark]

2 Denise likes to skateboard. In one particular trick, she runs and jumps onto the skateboard. Denise has a mass of 45 kg and her skateboard has a mass of 6.0 kg.

a Denise runs at a velocity of 4.0 m/s. What is her momentum?

.... [1 mark]

b Initially, the skateboard is stationary. What is its initial momentum?

.... [1 mark]

c When Denise jumps onto the skateboard, both she and the skateboard move off together. What is their new velocity? Show your working.

....

....

....

.... [3 marks]

3 Two identical railway carriages move towards each other at velocities of 2.0 m/s and 3.0 m/s respectively. They collide and couple together.

What is their combined direction and velocity after the collision?

.... [2 marks]

4 a Crash barriers are deliberately designed to crumple if a car collides with them. Explain how this helps protect the passengers in the car.

....

....

.... [2 marks]

b Name **two** other safety features in cars that are designed to reduce the force of an impact.

....

.... [2 marks]

5 A toy crossbow uses an elastic strap to apply force to an arrow. It applies an average force of 40 N to the arrow for 0.4 s.

What is the change in momentum of the arrow?

....

.... [2 marks]

Work, energy and power

1 a Describe what is meant by 1 joule of work.

....

.... [2 marks]

b A force of 20 N is exerted on a box as it moves through a horizontal distance of 5 m. Calculate the work done.

.... [1 mark]

c What work is done against gravity when a person of mass 60 kg rises a vertical height of 4 m in a lift?

.... [1 mark]

2 A rollercoaster car is pulled up a slope by a motor. The height of the slope is 8 m and the mass of the car is 500 kg.

a What work is done in pulling the car up the slope?

.... [1 mark]

b The motor output is 6000 J as the car is pulled up the slope. Explain why this is greater than your previous answer.

.... [1 mark]

c It takes 40 seconds for the car to reach the top of the slope. Find the power of the motor.

.... [1 mark]

3 A forklift truck lifts a pallet of bricks of weight 2400 N through a vertical height of 1 m in 40 seconds.

a Find the work done.

.... [1 mark]

b Find the power used by the forklift.

.... [1 mark]

4 A remote-controlled car moves at a constant velocity of 10 m/s. The average driving force of its motor is 2000 N.

What is the output power of the car?

.... [1 mark]

KE, GPE and conservation of energy

1 Complete these sentences:

  a Kinetic energy is measured in .... **[1 mark]**

  b Mass is measured in .... **[1 mark]**

  c Velocity is measured in .... **[1 mark]**

2 a What is the kinetic energy of a remote-controlled car of 12.0 kg travelling at 2.5 m/s?

  .... **[1 mark]**

  b Give **two** ways that the kinetic energy of the car could be increased.

  ....

  .... **[2 marks]**

---

3 What is the gain in gravitational potential energy of a climber of mass 50 kg who climbs a crag 200 m high?

  .... **[1 mark]**

4 Mathew is on his roller blades. He skates up a slope that is 18 m high and 40 m long. Mathew's mass is 30 kg.

  a What is Matthew's gain in gravitational potential energy?

  .... **[1 mark]**

  b He then skates back down the slope. What happens to the gravitational potential energy?

  .... **[1 mark]**

  c Find Matthew's maximum velocity down the slope, if he simply holds his feet still and allows himself to slide down freely.

  .... **[1 mark]**

  d In reality, when Mathew slides down freely, his maximum velocity is less than this. What has happened to some of the energy?

  .... **[2 marks]**

  e If Mathew skates down, he can achieve a higher velocity than this. What force allows him to do this?

  .... **[1 mark]**

5 A skydiver falls a distance of 500 m. Assuming no energy is lost, what is her gain in kinetic energy?

  .... **[1 mark]**

---

6 A car has a braking distance of 12 m when it is travelling at 10 m/s.

  What will happen to the braking distance if the velocity of the car is doubled? Explain your answer.

  ....

  ....

  .... **[3 marks]**

G–E

D–C

B–A*

Atomic nuclei and radioactivity

1 Scientists believe that everything in the Universe is made of atoms.

a Name **two** particles that are found inside the nucleus of an atom.

.... [2 marks]

b A particular atom has a neutral charge. What does this reveal about its protons and electrons?

.... [1 mark]

c What name is given to a particle with more electrons than protons?

.... [1 mark]

d What charge will this particle have?

.... [1 mark]

e Suggest a way in which charged atoms can be produced.

.... [1 mark]

G–E

2 The symbol below is used to represent an atom.

$$^A_Z X$$

a What does X represent? .... [1 mark]

b What does A represent? .... [1 mark]

c What does Z represent? .... [1 mark]

d How could you work out how many neutrons are inside the atom?

.... [1 mark]

D–C

3 What is an isotope?

....

....

.... [4 marks]

B–A*

4 There are three types of nuclear radiation.

a Which name is given to an electron emitted from the nucleus?

.... [1 mark]

b Which type of radiation is a helium nucleus?

.... [1 mark]

c Which type of radiation is an electromagnetic wave?

.... [1 mark]

G–E

5 Complete the parts **a** to **d** of the table below.

Radiation	Charge	Mass	Ionising effect	Penetration
Alpha	+2	a	Strong	b
Beta	c	0.00055	Weak	Stopped by a few millimetres of aluminium
Gamma	0	0	d	Never completely stopped, but reduced significantly by thick lead or concrete

[4 marks]

D–C

6 Explain what happens to the numbers of nucleons in an atom when it emits an alpha particle.

....

.... [2 marks]

B–A*

Nuclear fission

1 Nuclear reactions produce energy.

a What word describes the nuclear reaction that is responsible for the energy generated in the Sun?

.... [1 mark]

b What word is used to describe 'splitting the nucleus'?

.... [1 mark]

c What particle causes this nuclear reaction?

.... [1 mark]

2 What name is given to the process by which fast-moving particles from one reaction go on to split other uranium nuclei?

.... [1 mark]

3 a What **two** fuels are commonly used in nuclear power stations?

.... [2 marks]

b Explain why the disposal of waste from nuclear power stations is a major concern.

....

.... [2 marks]

4 The diagram below shows a nuclear reactor.

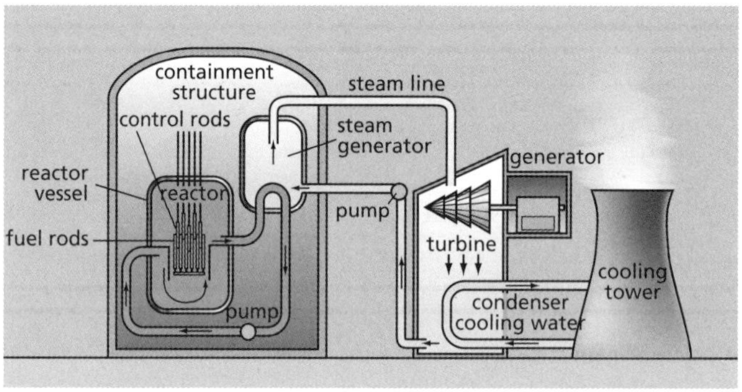

Explain the purpose of each of the following key components.

a Coolant

.... [1 mark]

b Moderator

....

.... [2 marks]

c Control rods

....

.... [2 marks]

5 Explain the meaning of the term 'critical mass'.

....

.... [2 marks]

Fusion on the Earth

1 The diagram below shows the process of nuclear fusion.

Use the words below to label the diagram.

helium   tritium   deuterium   neutron

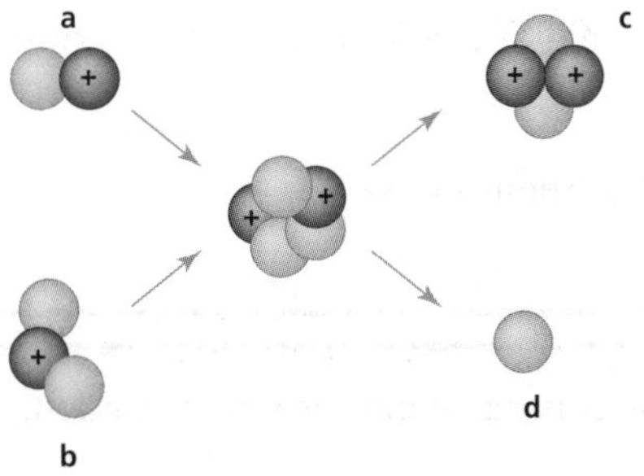

G–E

a

c

b

d

[4 marks]

2 The theory of cold fusion was proposed by scientists Stanley Pons and Martin Fleishmann in 1989.

a What is cold fusion?

....

[1 mark]

b Explain **two** reasons why the theory of cold fusion has been rejected.

D–C

....

....

....

[2 marks]

3 Fusion reactions are more difficult to start than fission reactions.

a What triggers a fission reaction?

....

[1 mark]

b Fusion reactions happen when two nuclei collide. Why must this collision take place at extremely high pressures and temperatures?

....

....

[2 marks] B–A*

c Describe how these nuclei can be controlled at such high temperatures.

....

....

[2 marks]

d Why is it possible for fusion to occur in stars?

....

....

[1 mark]

1 What name is given to the device that detects radiation?

.... **[1 mark]**

G–E

2 Background radiation comes from a variety of sources and is monitored in different regions of the UK.

a What type of rock contains uranium that can be a source of background radiation?

.... **[1 mark]**

b What radioactive gas is given off by these rocks?

.... **[1 mark]**

3 The pie chart below shows the main sources of background radiation.

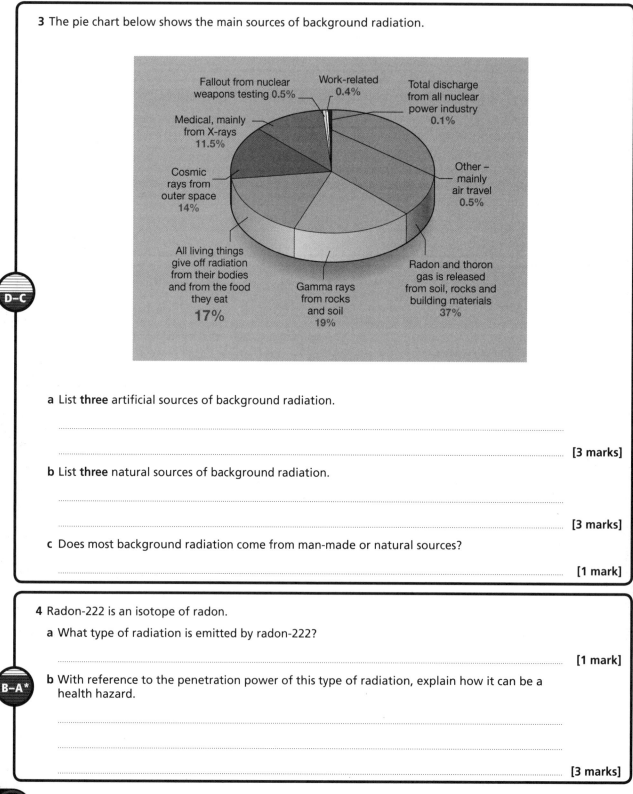

Fallout from nuclear weapons testing 0.5%

Work-related 0.4%

Total discharge from all nuclear power industry 0.1%

Medical, mainly from X-rays 11.5%

Other – mainly air travel 0.5%

Cosmic rays from outer space 14%

All living things give off radiation from their bodies and from the food they eat 17%

Gamma rays from rocks and soil 19%

Radon and thoron gas is released from soil, rocks and building materials 37%

D–C

a List **three** artificial sources of background radiation.

....

.... **[3 marks]**

b List **three** natural sources of background radiation.

....

.... **[3 marks]**

c Does most background radiation come from man-made or natural sources?

.... **[1 mark]**

4 Radon-222 is an isotope of radon.

a What type of radiation is emitted by radon-222?

.... **[1 mark]**

B–A*

b With reference to the penetration power of this type of radiation, explain how it can be a health hazard.

....

....

.... **[3 marks]**

Uses of radioactivity

1 Suggest **two** uses of radiation in industry.

....

.... [2 marks]

G–E

2 The diagram below shows a paper-making machine in a factory. The thickness of the paper is controlled by a radiation source.

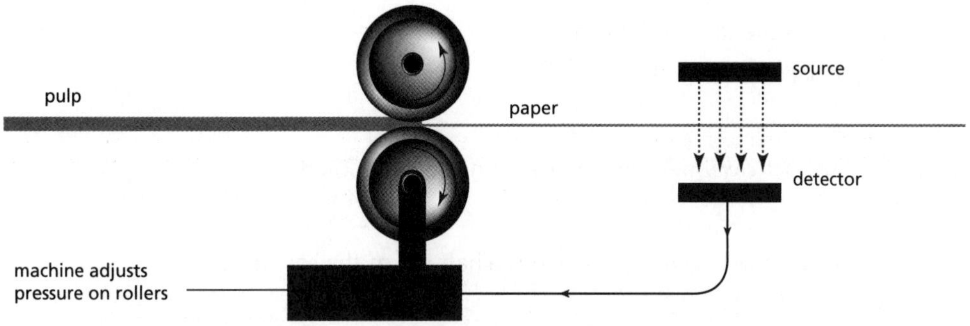

pulp

paper

source

detector

machine adjusts pressure on rollers

D–C

a What type of radiation source would be suitable for this job?

.... [1 mark]

b Describe how the detector signal changes if the paper is too thick.

.... [1 mark]

c Describe how the detector signal changes if the paper is too thin.

[1 mark]

3 In an electric smoke alarm, an alpha source is used.

Explain why an alpha source is the most suitable for this job.

....

....

.... [3 marks]

B–A*

4 What is the main advantage of using gamma rays to sterilise hospital equipment?

....

.... [1 mark]

G–E

5 Technetium-99m is a versatile tracer used in hospitals. It emits gamma rays and has a half-life of about 6 hours. The phrases below describe how it is used.

Put the sentences in the correct order.

i The tracer is carried around the body by the blood. It builds up in the cancerous regions of the patient's body.

ii A gamma camera is used to detect and display the gamma rays that pass through the patient.

iii The patient is injected with a small amount of radioactive tracer.

[3 marks]

D–C

Activity and half-life

1 Radioactive decay is random and spontaneous.

a What is meant by random?

.... [1 mark]

b What is meant by spontaneous?

.... [1 mark]

2 The activity of a source is the rate of decay of nuclei.

a What is the unit of activity?

.... [1 mark]

b What happens to the activity when the number of nuclei is doubled?

.... [1 mark]

c Describe the relationship between activity and the half-life of the source.

.... [1 mark]

3 A radioactive sample has a half-life of 6 hours. Initially the activity is 800 Bq. What will the activity be after:

a 12 hours?

.... [1 mark]

b 24 hours?

.... [1 mark]

4 State **three** safety precautions that should be used when handling radioactive sources in a laboratory.

....

....

.... [3 marks]

5 Patricia believes that nuclear power is better for the environment than coal because nuclear reactors do not give off any CO_2 emissions. Brian believes that nuclear power stations are worse for the environment.

Explain why Brian may be right.

....

.... [2 marks]

6 Describe **two** ways in which high-level waste from nuclear reactors can be disposed of.

....

.... [2 marks]

Topic 6: 6.4, 6.5, 6.6, 6.7, 6.8, 6.9, 6.10, 6.11, 6.12

P2 Extended response question

*Aran is given a radioactive source, suitable for use in a school laboratory.

Plan an experiment that Aran could use to decide whether the source emits alpha, beta or gamma radiation. You should describe any apparatus that is needed and explain how the results will determine the type of source.

[6 marks]

Questions labelled with **asterisk** (*) are ones where the quality of your written communication will be assessed. You should take particular care with your spelling, punctuation and grammar, as well as the clarity of expression in these answers.

Intensity of radiation

1 Some types of radiation are ionising and some are non-ionising. In the table below indicate the type of radiation. (The first line has been completed for you.)

Type of radiation	Ionising	Non-ionising
X-rays	✓	
Light waves		
Ultrasound		
Ultraviolet		
Microwaves		

[4 marks]

2 James and Joshua are investigating light intensity from an LED torch at different distances. They use a photographer's light meter to detect the light intensity at different distances in a darkened sports hall. The table to the right shows the results they obtained.

Distance (m)	Intensity
0.5	400
0.7	202
1.0	98
1.5	46
2	25
3	10

a Explain why they need to use a darkened hall to carry out the experiment.

....

....

....

.... [1 mark]

b Plot the results from the experiment on the graph paper provided and draw a curve of best fit through the points. **[3 marks]**

c Explain the shape of the graph.

....

....

....

....

.... [2 marks]

d Sketch another line on the graph showing the results they would expect if they repeated the experiment when the hall was full of smoke. **[1 mark]**

3 A communications aerial emits short-wave radio waves at a power of 600 W. A receiver can only detect short-wave radio at intensities of more than 0.5 W/m^2.

a Calculate the maximum working distance away from the mast.

....

....

....

.... [4 marks]

b Explain why the maximum working distance will in practice be less than the calculated value.

....

....

.... [2 marks]

Topic 1: 1.1d, 1.2, 1.3, 1.4

Properties of lenses

1 The diagrams below show three different lenses. For each one state whether it is a diverging lens or a converging lens.

a ....

b ....

c ....

[3 marks]

2 a The lens shown in fig **1a** above has a focal length of 15 cm. Calculate the power of the lens.

....

....

[2 marks]

b The lens shown in fig **1b** above has a focal length of 12 cm. Calculate the power of this lens.

....

....

[2 marks]

3 An optometrist uses many different lenses to make up the correct prescription for correcting a patient's eyes. The power of five of the lenses he wishes to use are as follows:

Lens	Power
A	+ 10
B	+ 3
C	+ 2
D	− 2
E	− 4

a If he combines lens A and lens E, calculate the focal length of the combined lenses.

....

....

[2 marks]

b The optometrist wants a lens combination with a focal length of 8.3 cm. Which two lenses should he use?

....

....

[2 marks]

c Now he wants a combination with a focal length of −0.5 m. Which lenses should he use?

....

....

[2 marks]

Lens equation

1 Katie and Lyra are trying to use a converging lens to create an image of a filament bulb. They use a converging lens with a focal length of 15 cm and place it 40 cm from the bulb. They use a cardboard screen to find the image.

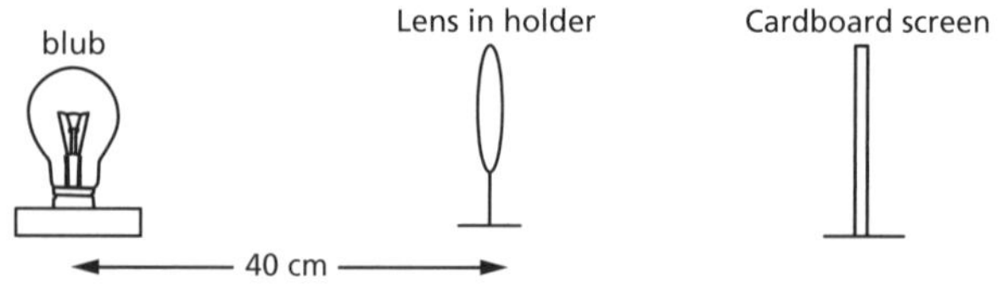

blub   Lens in holder   Cardboard screen

←—— 40 cm ——→

a Explain how they can measure the distance between the focused image and the bulb.

....

....

.... [2 marks]

b Circle the words which can be used to describe the image they obtain:

Virtual   Diminished   Inverted   Real [2 marks]

c What would happen to the image if they moved the lens so that it was only 20 cm from the bulb?

....

.... [2 marks]

2 Ray diagrams can be used to work out the position and size of an image formed with a lens. The ray diagram shows a converging lens of focal length 5 cm and the object is 3 cm high and is 12 cm from the lens.

Draw two rays of light from the object to find out the position and size of the image.

F   F

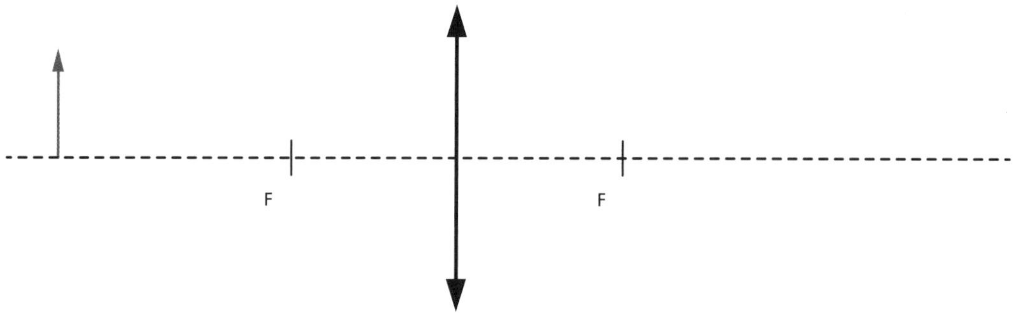

[5 marks]

3 A converging lens is used to project an image of a bulb 4 cm away from the lens. The focused image is 2 m away from the lens.

a Calculate the focal length of the lens.

....

....

.... [3 marks]

b The same lens is now being used as a magnifying glass. If the object to be magnified is 3 cm away from the lens, how far away from the lens will the image of the object appear?

....

....

.... [3 marks]

The eye

1 The diagram below shows a cross-section of the human eye.

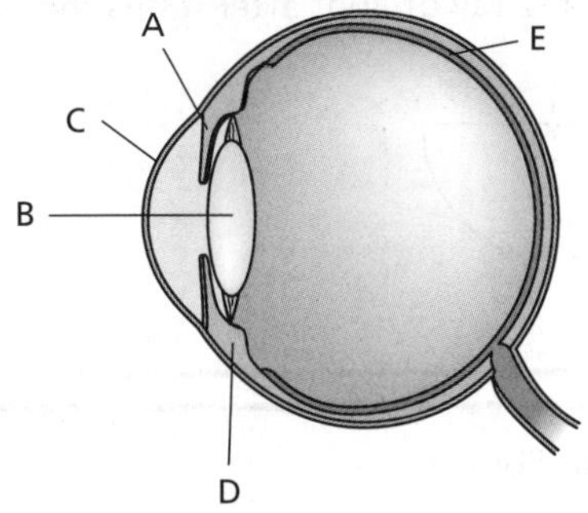

G–E

a Some parts of the eye are labelled with letters. Answer the following with the letter.

  i Which parts of the eye refract light? .... [1 mark]

  ii On which part of the eye is the image formed? .... [1 mark]

b Explain the function of part D.

....

.... [2 marks]

---

2 Explain what is meant by the terms 'near point' and 'far point' with respect to sight.

....

....

.... [3 marks]

3 The diagram below shows some rays of light entering an eye.

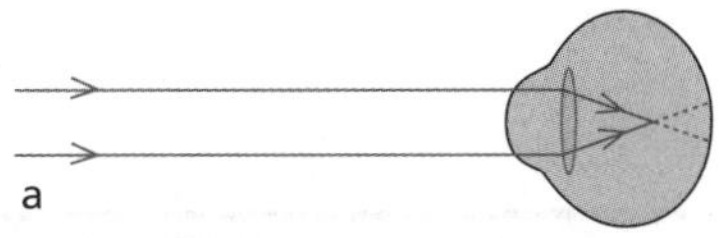

a

D–C

a Is this eye long-sighted or short-sighted?

.... [1 mark]

b What type of lens should be used to correct this defect?

.... [1 mark]

c The corrective lens can be supplied in either spectacles or contact lenses. Explain why someone might choose to use contact lenses instead of spectacles to correct their sight.

....

.... [2 marks]

4 Another more costly alternative is laser eye surgery. Describe one advantage and one disadvantage of using this technique to correct someone's vision.

B–A*

....

....

.... [2 marks]

Total internal reflection and endoscopes

1 The diagrams below show some rays of light entering a semi-circular glass block. Complete the diagrams to show what happens to the ray of light after it hits the flat surface of the block.

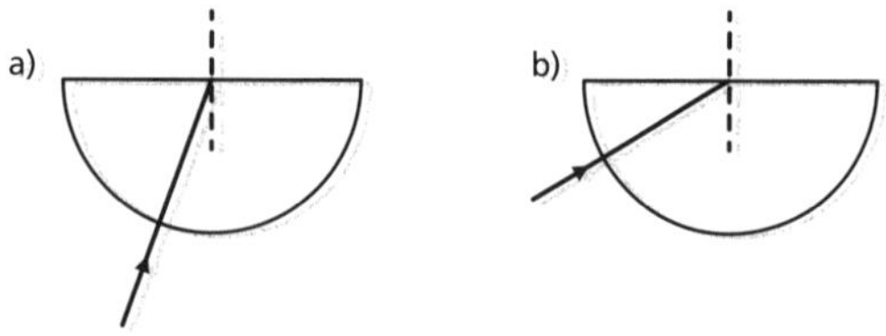

G-E

a)

b)

[4 marks]

2 The diagram below shows an optical fibre.

a Complete the path of the ray of light.

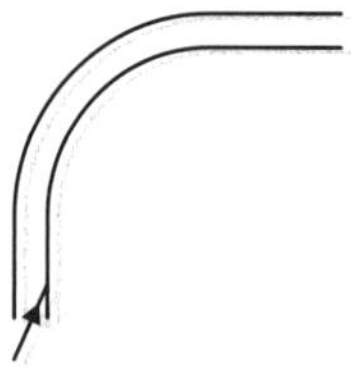

[2 marks]

D-C

b Optical fibres are flexible and can be bent around corners, but there is a minimum radius of curvature. If the fibre is coiled more tightly than this, it will no longer transmit the light. Using your knowledge of total internal reflection, explain why this is the case.

....

....

.... **[2 marks]**

c Optical fibres are used in endoscopes. What is an endoscope?

....

....

.... **[2 marks]**

3 The table below shows the refractive index and critical angle for some transparent materials.

a Complete the blanks in the table.

Material	Refractive index	Critical angle
Water		49°
Glass	1.50	
Diamond	2.35	
Perspex		43°

[4 marks]

B-A*

b The diagram shows light leaving a glass block into air.

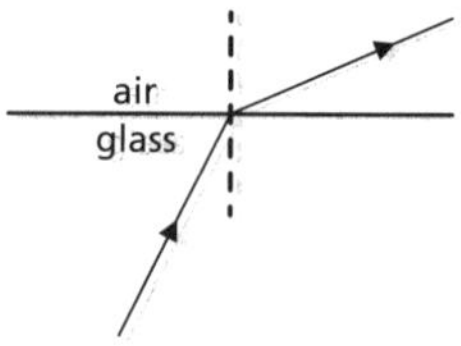

The light refracts because the speed of light is greater in air than it is in glass. This question is about what will happen if the glass block is placed in water.

air
glass

i Will the difference in the speed of light in the two materials be larger or smaller?

.... **[1 mark]**

ii Sketch a line on the diagram to show where the light ray will be refracted if it passes from glass to water. **[1 mark]**

iii What will happen to the critical angle? Explain your answer.

....

.... **[2 marks]**

Topic 1: 1.1c, 1.15, 1.16, 1.17, 1.18, 1.19, 1.20, 1.21

Medical uses of ultrasound

1 What is ultrasound?

.... **[2 marks]**

2 Ultrasound scans are often used to check the development of a foetus inside a mother's womb. Explain why X-rays are not used for this.

....

.... **[2 marks]**

3 During an ultrasound scan the size of the foetus's head is measured. The radiographer sends pulses of ultrasound towards the head and records the time the pulses take to return to the detector.

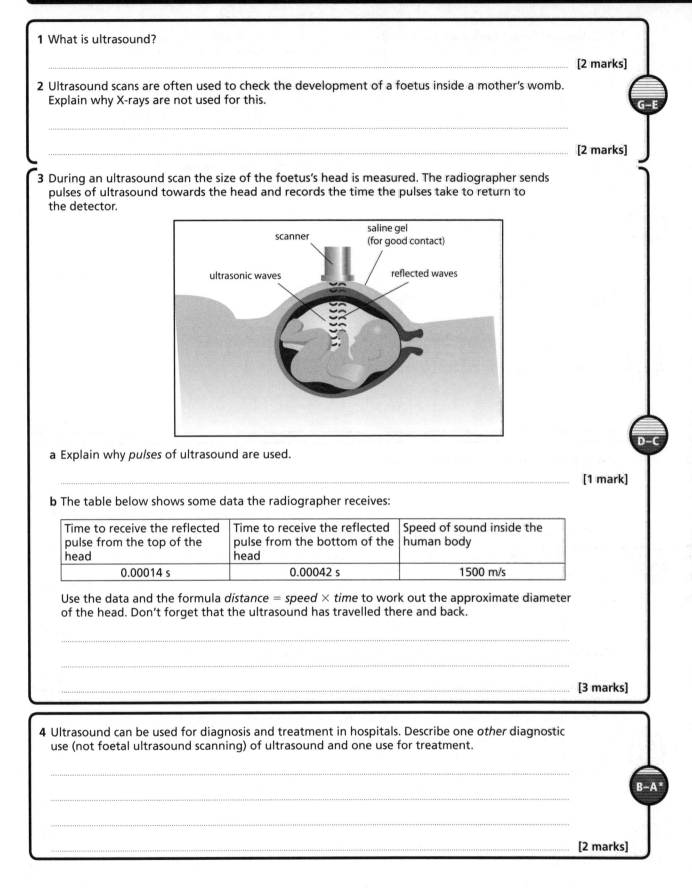

scanner

saline gel
(for good contact)

ultrasonic waves

reflected waves

a Explain why *pulses* of ultrasound are used.

.... **[1 mark]**

b The table below shows some data the radiographer receives:

Time to receive the reflected pulse from the top of the head	Time to receive the reflected pulse from the bottom of the head	Speed of sound inside the human body
0.00014 s	0.00042 s	1500 m/s

Use the data and the formula *distance = speed × time* to work out the approximate diameter of the head. Don't forget that the ultrasound has travelled there and back.

....

....

.... **[3 marks]**

4 Ultrasound can be used for diagnosis and treatment in hospitals. Describe one *other* diagnostic use (not foetal ultrasound scanning) of ultrasound and one use for treatment.

....

....

....

.... **[2 marks]**

Producing X-rays

1 This diagram shows a simple X-ray tube.

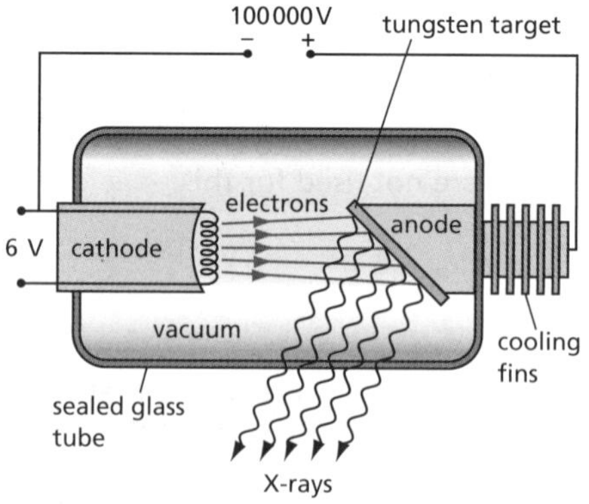

G–E

a Explain why the tube contains a vacuum.

....

.... [2 marks]

b What is the purpose of the low voltage (6 V) supply?

....

.... [2 marks]

c What is the purpose of the high voltage (100 000 V) supply?

....

.... [2 marks]

d Why does the anode need cooling fins?

....

.... [2 marks]

D–C

e Explain why the beam of electrons is equivalent to an electric current.

....

.... [2 marks]

f Explain how you could produce X-rays with a greater intensity.

.... [1 mark]

g Explain how you could produce X-rays with a greater frequency.

.... [1 mark]

2 An X-ray tube has an accelerating voltage of 65 000 V, with a current of 1.5 mA.

The charge on one electron $e = 1.6 \times 10^{-19}$ C and the mass of one electron $m_e = 9.1 \times 10^{-31}$ kg.

Calculate:

a The kinetic energy gained by an electron.

....

.... [2 marks]

B–A*

b The maximum speed with which the electron would strike the target.

....

.... [3 marks]

c The number of electrons arriving at the anode per second.

....

.... [2 marks]

Medical uses of X-rays

1 Casualties from a motor accident often have a CAT scan so that the doctors are able to assess any internal damage.

G–E

a Explain why a CAT scan is used instead of a conventional X-ray scan.

....

.... [2 marks]

b What are the risks associated with having a CAT scan?

....

.... [2 marks]

2 Fluoroscopy is a medical technique used in the diagnosis of digestive disorders. Describe how the technique can help doctors to find out the location of a blocked intestine.

....

....

....

G–E

.... [4 marks]

3 The intensity of X-rays follows an inverse square law with distance.

a Complete the table below to show the intensity of X-rays passing through lead.

Thickness of lead (mm)	1	2	3	4	5
Relative intensity of X-rays	200				

[2 marks] D–C

b Why do radiographers carrying out X-ray scans wear lead aprons?

....

.... [2 marks]

c Peter and Stephanie are discussing the treatment of cancer.
Peter thinks it is wrong to use X-rays to treat cancer because X-rays can also cause cancer.
Stephanie argues that the X-rays only kill the cancerous cells.

Discuss whether you agree or disagree with Peter and Stephanie.

....

....

B–A*

....

....

.... [4 marks]

Heart action and ECG

1 Jennifer measures her pulse rate. She counts 36 beats in 30 seconds.

a What is her pulse rate in beats per minute?

....

....

.... [1 mark]

b What is the frequency of her heart in Hertz?

.... [2 marks]

2 The diagram below shows an ECG trace. On the diagram label the parts of the curve which correspond to:

a When the atria contract [1 mark]

b When the ventricles contract [1 mark]

c When the ventricles relax [1 mark]

3 George has an irregular heartbeat and his doctor wishes to take an ECG of his heart.

a Describe how an ECG will be taken.

....

....

.... [3 marks]

b Describe what the doctor expects to see on the ECG trace.

....

.... [2 marks]

c Suggest the type of treatment which might be offered to George.

....

.... [2 marks]

Pulse oximetry

1 What two things are monitored by a pulse oximeter?

....

.... **[2 marks]**

2 The diagram below shows how a pulse oximeter works:

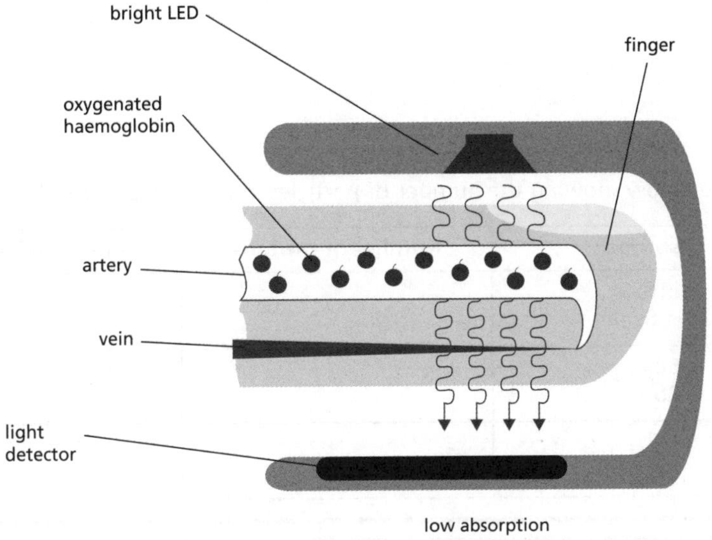

bright LED

finger

oxygenated
haemoglobin

artery

vein

light
detector

low absorption

a What is haemoglobin?

....

....

.... **[3 marks]**

b Explain how the reading on the light detector is related to the amount of oxygen in the blood

....

.... **[2 marks]**

c Why is pulse oximetry used to monitor patients in intensive care?

....

....

.... **[3 marks]**

d Explain why pulse oximeters use the absorption of two different wavelengths of light in order
to determine the percentage of oxygenated blood.

....

....

....

.... **[4 marks]**

G–E

D–C

B–A*

Topic 2.14

Unit P3 137

Ionising radiations

1 Explain the following terms:

a nucleon

....

.... [2 marks]

b isotope

....

....

.... [3 marks]

G–E

2 Complete the table below showing the number of particles in the isotopes listed.

	Isotope	Number of protons	Number of neutrons
a	$^{16}_{8}O$		
b	$^{54}_{26}Fe$		
c	$^{236}_{92}U$		

[6 marks]

3 Identify the type (or types) of nuclear radiation which is described in the statements below. Choose your answer from the following (some may be used more than once):

Alpha   Beta-minus   Beta-plus   Gamma   Neutron

a This type of radiation is the most highly ionising:

.... [1 mark]

b These types of radiation are positively charged:

.... [2 marks]

D–C

c These types of radiation are neutral:

.... [2 marks]

d This type of radiation is an anti-particle:

.... [1 mark]

e This type of radiation is a fast-moving electron:

.... [1 mark]

4 Medical professionals who work with nuclear radiation often wear a dosimeter. Explain the function of a dosimeter and how it is used to protect medical workers

....

....

....

....

.... [4 marks]

B–A*

Radioactive decays

1 The following unstable nuclei are alpha emitters. For each one, determine the daughter nucleus.

a $^{220}_{86}Rn$

.... [2 marks]

b $^{230}_{90}Th$

.... [2 marks]

c $^{238}_{92}U$

.... [2 marks]

2 Iodine-129 is an unstable isotope of iodine ($^{129}_{53}I$) which emits beta-minus particles.

a Explain how a nucleus, containing protons and neutrons, emits a beta-minus particle (electron).

....

....

.... [3 marks]

b What is the daughter nucleus following the decay of iodine-129?

.... [2 marks]

c Why are gamma rays also emitted during this decay?

.... [1 mark]

3 Write the nuclear decay equations for the following decays:

a Alpha decay of americium-241 $^{241}_{95}Am$

....

.... [3 marks]

b Beta-plus decay of potassium-40 $^{40}_{19}K$

....

.... [3 marks]

c Beta-minus decay of caesium-137 $^{137}_{55}Cs$

....

.... [3 marks]

Stability of nuclei

1 The N–Z graph is a plot of the atomic number N against the mass number of all isotopes. The curve shows the stable nuclei.

curve of stable nuclei

N (neutron number)

A (atomic number)

B–A*

a Explain the shape of the curve with reference to the strong nuclear force.

....

....

....

.... **[4 marks]**

b Mark a point B which would indicate a beta-minus emitter. **[1 mark]**

c Mark a point C which would indicate a beta-plus emitter. **[1 mark]**

d Mark a point D which would indicate an alpha emitter. **[1 mark]**

2 Explain why a neutron-poor isotope is a beta-plus emitter.

....

....

....

.... **[4 marks]**

3 What is a neutrino?

....

....

.... **[3 marks]**

Topic 3: 3.9, 3.10, 3.11, 3.12, 3.13.

Quarks

1 In the 1930s scientists believed that electrons, neutrons and protons were the fundamental particles which made up all matter.

However, now scientists have discovered that this is not true.

a Explain what is meant by the term 'fundamental particle'.

....

....

.... **[2 marks]**

b What do scientists now believe are the fundamental particles?

.... **[1 mark]**

2 Protons and neutrons are now believed to be made up of 3 quarks each.

a Describe the difference between protons and neutrons in terms of quarks.

....

....

.... **[2 marks]**

b Explain how a proton has an overall charge of +1e.

....

....

.... **[3 marks]**

3 Explain both types of beta decay in terms of quarks.

....

....

....

....

....

.... **[4 marks]**

Dangers of ionising radiations

1 Ionising radiations can damage living cells.

a Give three examples of symptoms of radiation damage.

....

....

....

.... [3 marks]

b Describe two methods of protecting people who work with radioactive materials from radiation damage.

....

.... [2 marks]

c Describe how radiation damage may lead to cancers.

....

....

.... [3 marks]

d Explain three factors which will affect the amount of cell damage caused by ionising radiation.

....

....

.... [3 marks]

2 The maximum permissible equivalent dose for medical personnel is 20 mSv per year averaged over 5 years, with a maximum of 50 mSv in any one year.

a What is meant by the term 'equivalent dose'?

....

.... [2 marks]

b The radiation dose for a brain CAT scan is 5 mSv.

Discuss whether this type of treatment is safe for the patient and the radiographer.

....

....

....

.... [4 marks]

Treatment of tumours

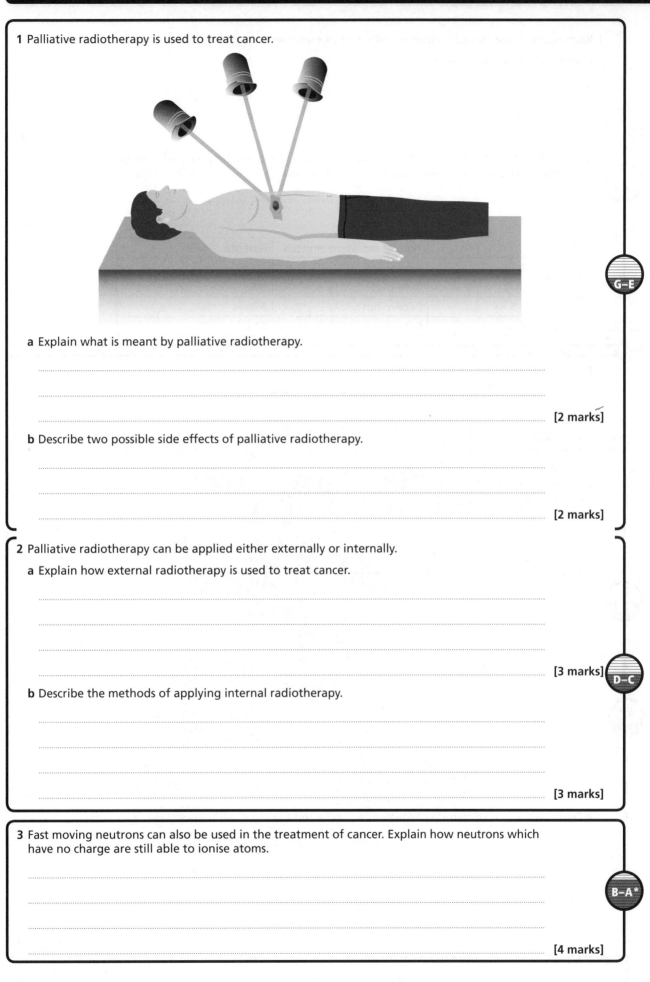

1 Palliative radiotherapy is used to treat cancer.

a Explain what is meant by palliative radiotherapy.

....

....

.... **[2 marks]**

b Describe two possible side effects of palliative radiotherapy.

....

....

.... **[2 marks]**

G–E

2 Palliative radiotherapy can be applied either externally or internally.

a Explain how external radiotherapy is used to treat cancer.

....

....

....

.... **[3 marks]**

b Describe the methods of applying internal radiotherapy.

....

....

....

.... **[3 marks]**

D–C

3 Fast moving neutrons can also be used in the treatment of cancer. Explain how neutrons which have no charge are still able to ionise atoms.

....

....

....

.... **[4 marks]**

B–A*

Diagnosis using radioactive substances

1 Radioactive tracers are often used to help doctors diagnose patients with many different ailments.

a What is a radioactive tracer?

....

....

.... [3 marks]

b Iodine is used as a radioactive tracer to detect problems in the kidney. The table below shows some different isotopes of iodine. Explain which isotope is the most suitable to use.

....

.... [3 marks]

Isotope	Type of radiation emitted	Half life
Iodine-108	Alpha	36 ms
Iodine-113	Beta-plus	6.6 s
Iodine-131	Beta-minus	8 days
Iodine-136	Beta-minus	83 s

2 Many hospitals in the developed world have a PET scanner.

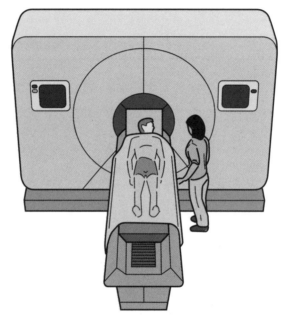

a PET scanners are used to monitor which type of condition?

....

.... [2 marks]

b Describe the main differences between PET scans and the use of ordinary radioactive tracers.

....

....

....

....

.... [4 marks]

Particle accelerators and collaboration

1 The Large Hadron Collider is the largest particle accelerator in the world.

a What is a 'particle accelerator'?

....

....

[2 marks]

b Explain why scientists need particle accelerators.

....

....

....

....

[2 marks]

G–E

c The LHC is a circular particle accelerator with a diameter of about 9 km. The particles make about 11 000 revolutions per second.

i Why is the LHC circular?

....

....
[2 marks]

D–C

ii How far could a particle travel in one second in the LHC?

....

....
[2 marks]

2 The LHC is run by CERN on the French–Swiss border. CERN is a European Organisation for Nuclear Research and is funded by the EU member states.

B–A*

a What are the advantages of international collaboration for nuclear research?

....

....

....

....

[2 marks]

b Discuss whether CERN should be a profit-making organisation.

....

....

....

....

[3 marks]

Cyclotrons

1 The diagram below shows a hammer-thrower spinning her hammer round in a circle.

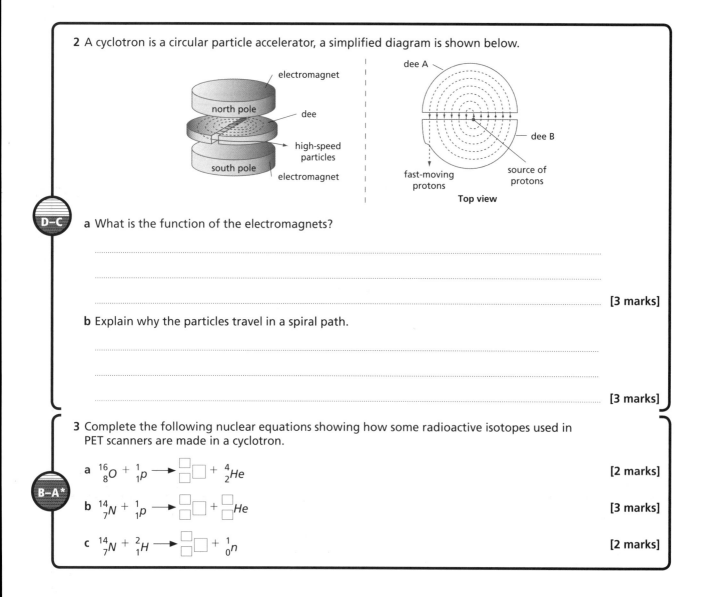

G–E

a Draw an arrow to show the direction of the velocity of the hammer – label it v. [1 mark]

b Draw an arrow to show the direction of the net force acting on the hammer – label it F. [1 mark]

c What is the name given to the force used to keep an object moving in a circle?

.... [1 mark]

2 A cyclotron is a circular particle accelerator, a simplified diagram is shown below.

D–C

a What is the function of the electromagnets?

....

....

.... [3 marks]

b Explain why the particles travel in a spiral path.

....

....

.... [3 marks]

3 Complete the following nuclear equations showing how some radioactive isotopes used in PET scanners are made in a cyclotron.

B–A*

a $^{16}_{8}O + ^{1}_{1}p \longrightarrow \square\square + ^{4}_{2}He$ [2 marks]

b $^{14}_{7}N + ^{1}_{1}p \longrightarrow \square\square + \square He$ [3 marks]

c $^{14}_{7}N + ^{2}_{1}H \longrightarrow \square\square + ^{1}_{0}n$ [2 marks]

Electron–positron annihilation

1 When an electron and a positron collide they annihilate each other and emit 2 gamma rays.

a Explain what happens to the masses of the electron and the positron.

....

.... [2 marks]

b Explain how momentum is conserved in the annihilation.

.... [2 marks]

c Explain how charge is conserved in the annihilation.

....

....

.... [3 marks]

2 Radioactive tracers which are positron emitters are tagged onto compounds, which are injected into a patient who will undergo a PET scan. The PET scanner consists of a ring of gamma ray detectors.

Explain how the location of the tagged compound can be identified.

....

....

....

.... [4 marks]

3 During some electron–positron annihilation 3.9×10^{-12} Joules of energy were emitted in the form of gamma rays.

Calculate the number of electron–positron pairs which were annihilated.
Mass of electron/positron = 9.11×10^{-31} kg.

....

....

.... [3 marks]

Momentum

1 Simon rolls a large ball towards a small ball so that they collide. After the collision the large ball moves off in a straight line as the small ball stops.

— 0 m/s →

— 10 m/s →

Before

— v →

— 0 m/s →

G–E

a If the mass of the small ball is 50 g, calculate its initial momentum.

....

.... **[2 marks]**

b If the mass of the small ball is 250 g, what is its initial momentum?

.... **[1 mark]**

c Use the principle of conservation of momentum to find out the final speed of the small ball.

....

....

.... **[3 marks]**

2 Collisions can be either elastic or inelastic.

a Explain what is meant by the terms elastic and inelastic collision.

....

.... **[2 marks]**

b For the example above calculate the difference in kinetic energy before and after the collision.

D–C

....

....

.... **[4 marks]**

c What has happened to this energy?

....

.... **[2 marks]**

3 A bullet is fired at a stationary block of wood with mass 0.5 kg. The bullet becomes embedded in the wood and moves off at a velocity of 1.3 m/s. If the mass of the bullet is 0.02 kg calculate the velocity of the bullet as it hit the piece of wood.

B–A*

....

....

.... **[3 marks]**

Topic 4: 4.7, 4.8, 4.9, 4.10, 4.11, 4.12

Matter and temperature

1 When an ice cube is placed on a surface at room temperature it melts. Explain what happens in terms of the water molecules.

....

....

.... **[3 marks]**

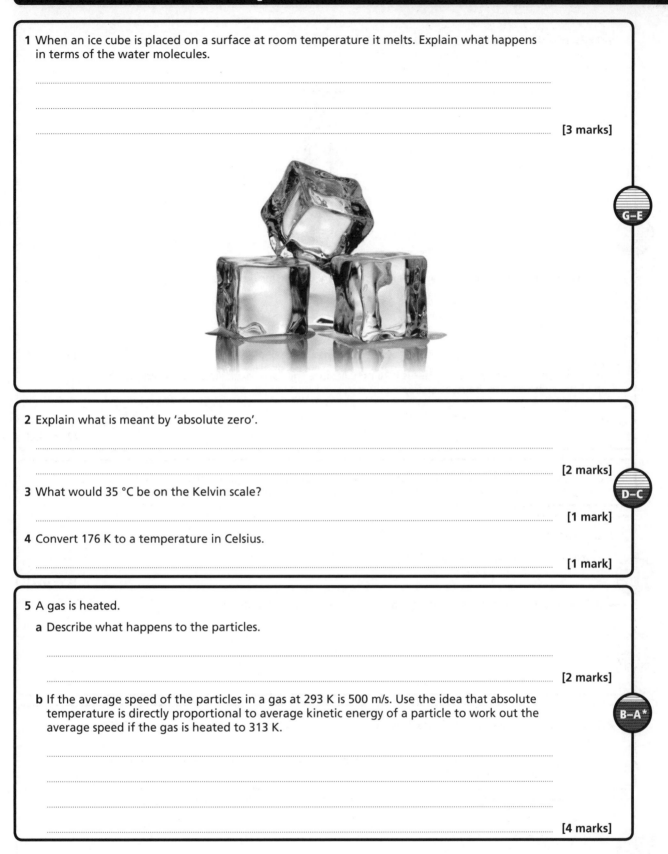

G–E

2 Explain what is meant by 'absolute zero'.

....

.... **[2 marks]**

3 What would 35 °C be on the Kelvin scale?

D–C

.... **[1 mark]**

4 Convert 176 K to a temperature in Celsius.

.... **[1 mark]**

5 A gas is heated.

  a Describe what happens to the particles.

....

.... **[2 marks]**

  b If the average speed of the particles in a gas at 293 K is 500 m/s. Use the idea that absolute temperature is directly proportional to average kinetic energy of a particle to work out the average speed if the gas is heated to 313 K.

B–A*

....

....

....

.... **[4 marks]**

Investigating gases

1 Oxygen is compressed and stored in a cylinder at a pressure of 320×10^5 Pa.

a Explain how the oxygen molecules exert a pressure on the inside walls of the cylinder.

....

....

.... **[3 marks]**

b If the surface area of the cylinder is 0.16 m² calculate the force exerted on the inside walls of the cylinder. Use the relationship F = P × A.

....

.... **[2 marks]**

c The oxygen cylinder has a volume of 2×10^{-3} m³. If the oxygen is transferred to another cylinder with volume 3×10^{-3} m³, calculate the new pressure, providing there is no change in temperature.

....

....

.... **[3 marks]**

d Explain using the kinetic theory of gases why the pressure changes.

....

....

.... **[3 marks]**

2 Some carbon dioxide is sealed into a syringe. The initial volume at room temperature (25 °C) is 7.0×10^{-6} m³.

a Explain using the kinetic theory of gases, what will happen to the volume of the gas if the temperature is increased to 70 °C.

....

....

....

.... **[3 marks]**

b Calculate the new volume of carbon dioxide.

....

....

.... **[4 marks]**

The gas equation

1 The diagram below shows some apparatus which can be used to find out how the pressure of a gas changes with temperature.

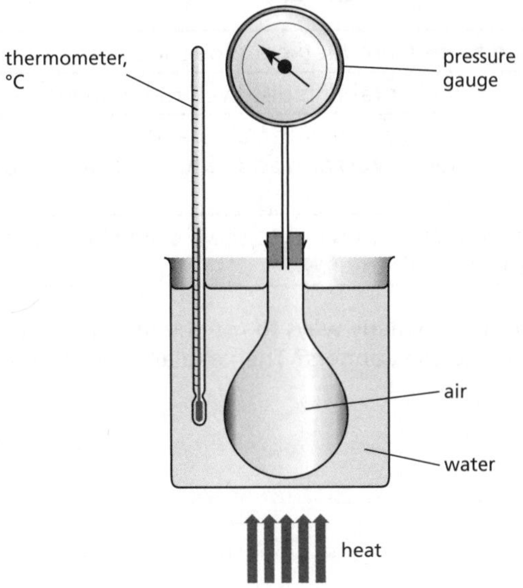

thermometer, °C

pressure gauge

air

water

heat

a Describe what needs to be measured.

....

.... **[2 marks]**

b Explain what results you would expect to get from the experiment.

....

....

.... **[3 marks]**

2 Which of the graphs below shows the relationship between the following properties of gases?

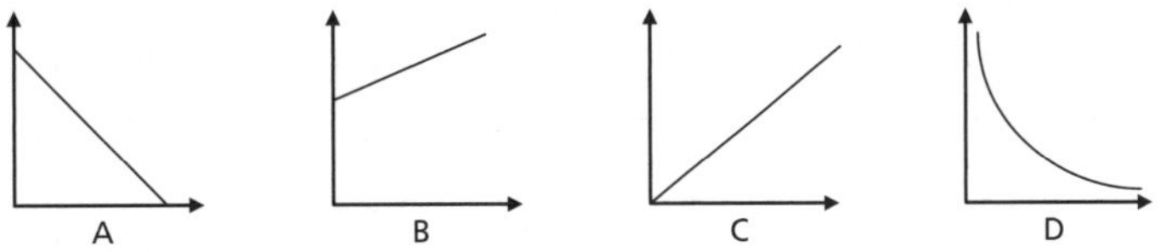

A   B   C   D

a Pressure of a gas and temperature of a gas in Kelvin   **[1 mark]**

b Volume of a gas and temperature of a gas in °Celsius   **[1 mark]**

c Pressure of a gas and volume of a gas   **[1 mark]**

3 A medical cylinder with volume 0.02 m³ contains some nitrogen at a pressure of 200×10^5 Pa. It is stored outside at 10 °C.

a Calculate the volume of nitrogen which would be released into the atmosphere at room temperature (25 °C). Atmospheric pressure is 1.0×10^5 Pa.

....

....

....

.... **[4 marks]**

b Calculate the time taken to empty the cylinder if the flow rate is 2.5 litres per minute (1 litre = 1×10^{-3} m³).

....

.... **[2 marks]**

*Radioactivity was discovered in 1896 by Henri Becquerel. Many scientists were very excited about the new discovery and devised many ways of using the 'new' radiation. The table below shows some of the uses of radioactivity for health during the first half of the 20th century.

Pedoscopes	X-rays used to measure feet in a shoe shop.
Spas	Bathing in water containing dissolved radon made people feel better.
Radon water	This was drunk by people with bad breath, insanity, tuberculosis and malnutrition. Radon was also added to some brands of chocolate, bread and toothpaste.
Hormesis	Experiments were carried out to see if a small amount of radiation could stimulate living cells to grow.

Why do you think people thought of so many ways to use the radioactivity in the early 20th century, and why they did not realise the dangers? Then explain the specific risks associated with these examples.

....

....

....

....

....

....

....

....

....

....

....

....

....

....

....

....

....

....

....

....

....

....

....

....

[6 marks]

Questions labelled with **asterisk** (*) are ones where the quality of your written communication will be assessed. You should take particular care with your spelling, punctuation and grammar, as well as the clarity of expression in these answers.

I know that ideas and models about the Solar System and the Universe have changed throughout history.	
I understand that waves are reflected and refracted at boundaries between different materials.	
I know that convex lenses are used in telescopes to magnify distant objects.	
I understand that waves transmit energy and information without transferring matter.	
I understand the terms frequency, wavelength and amplitude.	
I can describe the two types of wave – longitudinal and transverse.	
I can recall the electromagnetic waves in order of increasing wavelength and/or frequency.	
I understand the hazards of electromagnetic waves.	
I know uses for all the types of electromagnetic waves.	
I know that alpha, beta and gamma are types of ionising radiation, which come from radioactive materials.	
I can compare the relative sizes of and distances between the Earth, the Moon, the planets, the Solar System, galaxies and the Universe.	
I know that the Big Bang theory explains the origins of the Universe.	
I know what ultrasound and infrasound are.	
I know that there are two main types of seismic wave produced by earthquakes: P waves and S waves.	
I know that seismometers are used to detect earthquakes, but that earthquakes are almost impossible to predict.	
I can recall the different layers in the Earth's structure.	
I know that electric current (measured in amps) is a flow of charge and that the voltage pushes the current around a circuit.	
I know that power is the rate of using energy and is measured in watts.	
I can name some renewable energy sources and some non-renewable energy sources.	
I know that electricity is generated when a wire or a coil moves relative to a magnetic field.	
I know that mains electricity is a.c. and electricity from a battery is d.c.	
I know that electricity is transmitted at very high voltages and that it is dangerous.	
I know that the unit of electricity supply is the kilowatt hour.	
I understand the advantages of using low-energy appliances in the home.	
I know the different types of energy and know that energy is transferred from one type to another.	
I know that energy is often wasted in the form of heat.	
I know that the amount of heat (infrared) radiated by objects depends on the temperature and the surface.	
I am working at grades G/F/E	

I understand the differences and similarities between the geocentric model and the heliocentric model of the Universe.	
I understand how scientists use different waves to find out about the Universe.	
I know how to measure the focal length of a convex lens.	
I can explain how a reflecting telescope works.	
I can calculate the speed of a wave.	
I can describe uses for all types of electromagnetic waves.	
I understand ionisation and its dangers.	
I can describe methods used to find evidence of life beyond Earth.	
I understand how modern methods of astronomy have developed our understanding of the Universe.	
I can describe the evolution of stars.	
I can describe the evidence that supports both the Big Bang and the Steady State theories.	
I know that the Doppler shift occurs when a wave source is moving relative to an observer.	
I know that the red-shift of light from stars gives evidence of the Universe expanding.	
I can describe uses of both ultrasound and infrasound.	
I know that P waves are longitudinal and that S waves are transverse, and that they travel through the Earth at different speeds and are reflected and refracted at boundaries.	
I know that the Earth's crust is made up of tectonic plates, which move due to convection currents in the mantle.	
I understand that earthquakes are caused by the relative motion of the tectonic plates at their boundaries.	
I understand that electric current is the rate of flow of electric charge, and that the voltage or potential difference is a measure of the energy carried by the charge.	
I can calculate electrical power.	
I understand the advantages and disadvantages of using renewable and non-renewable sources of energy.	
I understand that ultimately the Sun is the source of all our energy on Earth.	
I understand how a generator produces electricity.	
I know that transformers are used to change the voltage of a.c. electricity.	
I understand that electricity is transmitted at very high voltages to reduce the amount of energy wasted.	
I can calculate electricity consumption in kilowatt hours and work out the cost.	
I understand the concept of payback time.	
I understand the principle of conservation of energy.	
I can calculate efficiency of energy transfers from data or from Sankey diagrams.	
I know that for an object to remain at constant temperature, the amount of energy absorbed must equal the amount of energy emitted.	
I am working at grade D/C	

I understand that refraction happens because the wave changes speed and direction at boundaries between different materials.	
I can interpret data about modern astronomical observations.	
I understand the evolution of stars.	
I understand how the wavelength and frequency change due to the Doppler effect.	
I understand that the red-shift of light from stars supports both the Big Bang theory and the Steady State theory, and the presence of cosmic background radiation only supports the Big Bang theory.	
I understand how P and S waves travel through the Earth.	
I can use the turns ratio equation for transformers to calculate relative potential differences across transformers.	
I can draw scaled Sankey diagrams to show the efficiency of energy transfers.	
I am working at grade B/A/A*	

I can recall that like charges repel and unlike charges attract.	
I can demonstrate an understanding of attraction by electrostatic induction and the dangers of electrostatic charges.	
I can recall that current is the rate of flow of charge.	
I can recall that cells and batteries supply direct current.	
I can describe how an ammeter is connected in series and measures current in amperes.	
I can understand that the current in a circuit can be changed using a variable resistor.	
I know the advantages and disadvantages of heating effect of an electric current.	
I can demonstrate an understanding of displacement and velocity.	
I can recall that velocity is speed in a stated direction and is a vector quantity.	
I can interpret distance–time and velocity–time graphs.	
I can draw and interpret free-body force diagrams.	
I can calculate resultant forces.	
I can use the equation $W = mg$.	
I can recall that stopping distance = thinking distance + braking distance.	
I can describe the factors that affect stopping distance.	
I can recall that momentum is a vector quantity.	
I can use the equation work done = force × distance moved in the direction of force.	
I can use the equation $KE = \frac{1}{2} mv^2$.	
I can state how atoms may gain or lose electrons to form ions.	
I can recall that alpha and beta particles and gamma rays are ionising radiations.	
I can describe nuclear fusion and fission reactions as a source of energy.	
I can describe how nuclear power stations get their energy from fission reactions and convert this into electricity.	
I can describe the process of nuclear fusion and recognise it as the energy source for stars.	
I can describe background radiation and the regional variation of radon gas.	
I can describe uses of radioactivity including irradiating food and sterilising equipment.	
I can recall how activity of a source decreases over a period of time.	
I can state that half-life is the time taken for half the undecayed nuclei to decay.	
I understand that ionising radiations damage tissues and can cause mutations.	
I can describe how scientists have changed their awareness of the hazards of radioactivity over time.	
I am working at grades G/F/E	

I can explain how an insulator can be charged by friction.	
I can explain electrostatic charging in terms of transfers of electrons.	
I understand common electrostatic phenomena in terms of movement of electrons.	
I can explain the uses of electrostatic charges, including paint and insecticide sprayers.	
I can explain how earthing safely removes excess charges.	
I can use the equation $Q = It$.	
I can explain that current is conserved at a junction.	
I can describe how a voltmeter is connected in parallel and measures potential difference in volts.	
I can use the equation $V = IR$.	
I understand how current varies with potential difference for filament lamps, diodes and fixed resistors.	
I can use the equations $P = VI$ and $E = IVt$.	
I can describe how the resistance of an LDR changes with light intensity and how the resistance of a thermistor changes with temperature.	
I can use the equation speed = distance / time.	
I can use the equation $a = (v-u) / t$.	
I can calculate acceleration from the gradient of a velocity–time graph.	
I can demonstrate an understanding of action and reaction forces.	
I understand how a body accelerates in the direction of a resultant force.	
I can use the equation $F = ma$.	
I can describe how bodies fall at the same acceleration in vacuum.	
I understand the relationship between weight and mass.	
I know how thinking and braking distance affect stopping distance.	
I can use the equation momentum = mass × velocity.	
I can describe the rate of change of momentum when applied to seat belts, crumple zones and air bags.	
I can describe power as the rate of doing work and can recall it is measured in watts.	
I can use the equation $GPE = mgh$.	
I can describe the idea of conservation of energy in energy transfers.	
I can describe an alpha particle as a helium nucleus, beta particle as an electron and gamma ray as electromagnetic radiation.	
I can compare alpha, beta and gamma radiations in terms of ionisation and abilities to penetrate.	
I can explain fission of uranium-235 and the principles of a controlled chain reaction.	
I can explain how a chain reaction is controlled in a nuclear reactor.	
I can describe how heat energy in a reactor is turned into electrical energy.	
I can explain the difference between nuclear fission and nuclear fusion.	
I can explain the origins of background radiation on the Earth.	
I can describe uses of radioactivity including smoke alarms, tracing and gauging thicknesses, and diagnosis and treatment of cancer.	
I can discuss the long-term disposal and storage of nuclear waste.	
I can evaluate the advantages and disadvantages of nuclear power for generating electricity.	
I am working at grades D/C	

I can explain that potential difference is energy transferred per unit charge.	
I can explain the transfer of heat in components in terms of collision between electrons and atoms in the lattice.	
I can calculate distance travelled from the area under a velocity–time graph.	
I can explain how resultant forces affect the motion of objects.	
I can describe the motion of objects falling in air.	
I can describe conservation of linear momentum.	
I can use and apply the equation force = rate of change of momentum.	
I can use the equation power = work done / time.	
I can use calculations to show that braking distance is directly proportional to the square of speed.	
I can describe the nuclei of isotopes.	
I can explain how atoms form ions.	
I can explain why nuclear fusion cannot occur at low pressures and temperatures and the difficulties of using it for power generation.	
I can use the concept of half-life to carry out calculations of the decay of nuclei.	
I am working at grades B/A/A*	

P3 Grade booster checklist

I know that the word radiation is used to describe any form of energy originating from a source, including both waves and particles.	
I understand that radiation can be ionising or non-ionising.	
I can describe how light is refracted through both converging and diverging lenses.	
I can identify the main parts of the eye, and understand their basic functions.	
I understand that light can be totally internally reflected when the angle is greater than a critical angle.	
I know that ultrasound is used in diagnosis.	
I know that X-rays are used in CAT scans and fluoroscopes.	
I can explain that a beam of electrons is equivalent to an electric current.	
I know that an ECG is a recording of the electrical activity of the heart.	
I can calculate frequency.	
I can describe the properties and nature of alpha, beta-minus and gamma radiation.	
I can describe the process of alpha decay.	
I can explain the precautions needed to ensure the safety of people exposed to radiation.	
I know that cancers can be treated with palliative radiotherapy.	
I can describe the use of radioactive tracers for diagnosis of some medical conditions.	
I know that scientists use particle accelerators to develop better understanding of the physical world.	
I can explain that a centripetal force is needed for an object to move in a circle.	
I know that if a positron collides with an electron they will both be annihilated and a gamma ray will be produced.	
I understand that momentum must be conserved in all collisions.	
I can describe the three states of matter in terms of the kinetic theory model.	
I can explain how the motion of particles in a gas causes pressure.	
I understand the relationship between the temperature of a gas and the speed of the particles.	
I can convert temperatures between the Celsius and Kelvin scales.	
I am working at grades G/E	

I understand that the intensity of radiation will depend on both the distance from its source and the nature of the medium through which it is travelling.	
I can relate the power of a lens to its shape, and know how to calculate it from the focal length.	
I can describe the symptoms and causes of both long sight and short sight, and know which type of lens is used to correct them.	
I understand how optical fibres and endoscopes make use of total internal reflection.	
I know that refraction occurs when light has a different speed in one medium from another.	
I understand how ultrasound can be used to create an image of the inside of a body.	
I understand that the ionisation by X-rays is related to its energy and hence frequency.	
I understand how X-rays are produced.	
I can describe how X-rays are used in CAT scans and fluoroscopes.	
I understand the inverse square law for electromagnetic radiation.	
I understand the risks and advantages of using X-rays for treatment and diagnosis.	
I can relate the shape of an ECG to heart action.	
I know that a pacemaker can be used to regulate heart action.	
I can describe how a pulse oximeter works.	
I can evaluate the social and ethical issues relating to the use of radioactive techniques in medical physics.	
I can describe the properties and nature of positron and neutron emission.	
I can describe the process of beta-minus and gamma decay.	
I know that many nuclei undergoing radioactive decay also emit gamma rays to lose excess energy.	
I can describe the dangers of ionising radiation in terms of tissue damage and possible mutations.	
I can compare and contrast the treatment of tumours using radiation applied internally and externally.	
I can explain the principles of using radioactive isotopes in PET scanners, and explain why the isotopes need to be produced nearby.	
I understand the reasons for collaborative, international research into big scientific questions such as particle physics.	
I can explain the basic principles of how a cyclotron accelerates charged particles in circular paths.	
I understand that particle accelerators are used to accelerate protons to be bombarded at stable isotopes to create radioactive isotopes for use in PET scanners.	
I know that PET scanners detect the gamma rays given off when a positron emitted by a radioactive isotope is annihilated in a collision with an electron.	
I understand that energy is only conserved in elastic collisions.	
I can explain the concept of absolute zero on the temperature scale.	
I understand that the average kinetic energy of the particles in a gas is directly proportional to the Kelvin temperature of the gas.	
I understand that the volume of a gas (at constant pressure) is directly proportional to its temperature in Kelvin.	
I understand that the pressure of a gas (at constant temperature) is inversely proportional to the volume of the gas.	
I understand that the pressure of a gas (at constant volume) is directly proportional to its temperature in Kelvin.	
I am working at grades D/C	

I can calculate the intensity of radiation in W/m^2.	
I know how to use the lens equation.	
I understand that laser eye correction can be used to permanently correct people's vision.	
I can use Snell's Law.	
I can describe ways that ultrasound is used for medical treatment.	
I can calculate the electric current flowing when a beam of electrons moves through an evacuated tube.	
I can calculate the kinetic energy of an electron after it has been accelerated through a potential difference.	
I can describe the process of beta-plus decay.	
I can write balanced nuclear equations to represent radioactive decay.	
I can describe the features of the N–Z curve for stable isotopes.	
I can identify the type of radioactive decay an unstable isotope will undergo from its position on the N–Z curve.	
I can describe the arrangement of up and down quarks in protons and neutrons.	
I can explain beta-minus and beta-plus decay in terms of quarks.	
I can use Einstein's conservation of mass-energy principle in a quantitative way.	
I can quantitatively use both the principle of conservation of momentum and the principle of conservation of energy for two body collisions in one dimension.	
I can use the equation $P_1 V_1 / T_1 = P_2 V_2 / T_2$.	
I am working at grades B/A*	

Notes

Notes

Answers

P1 Universal physics

Page 88 The Solar System

1 a Geocentric means that Earth is in the centre; and all the other objects orbit the Earth

b Mercury, Venus, Mars, Jupiter and Saturn (1 mark for two or three correct, 2 marks for four, 3 marks for all five)

c The ancient Greeks could only see five planets with the naked eye; nowadays we use telescopes to see further into space so can see more planets

d The retrograde motion of Mars / sometimes Mars appeared to be going backwards in space; the geocentric model explained this by using epicycles; the heliocentric model predicted this motion

2 Any three from: the Catholic Church believed that God had created the Universe; all the heavenly bodies were perfectly spherical and orbited Earth; if there were mountains on the Moon this meant that it wasn't a perfect sphere; if there were moons orbiting Jupiter then not everything could be orbiting Earth

3 Any two from: telescopes give brighter images; images are more magnified in a telescope; using a telescope enables astronomers to see more distant objects

4 The asteroid belt is made up of lumps of rock; orbiting the Sun between Mars and Jupiter

5 Any two from: the Hubble space telescope has a much larger aperture than ground-based telescopes; so it can see much dimmer / more distant objects; light doesn't have to pass through the atmosphere; so there is not as much distortion

Page 89 Reflection, refraction and lenses

1 Incidence; angle; reflected; normal

2
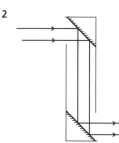
1 mark for right-angle turn at top mirror; 1 mark for right-angle turn at bottom mirror; 1 mark for arrows in correct direction (1 mark deducted if only one ray complete)

3 Diagram C is correct

4 Air; because it is the least dense material

5 a The focal point is the point at which parallel rays of light entering the lens; will converge and meet

b Lens A

Page 90 Lenses in telescopes

1 Obtain a focused image of a distant object; on a screen; and measure the distance from the image to the lens

2
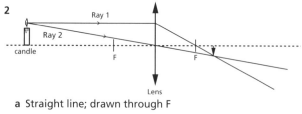

a Straight line; drawn through F

b Straight line; drawn through centre of the lens

c Image drawn where the rays cross

d Magnified

e Magnification = 0.5/1 = 0.5

f A real image can be projected; onto a screen and a virtual image cannot

3
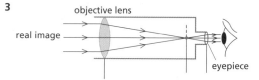

4 A diminished image of the distant object; is formed at the focal point of the objective lens; the eyepiece magnifies the image

5 It is a reflecting telescope; reflecting telescopes are lighter than refracting ones so it was easier to launch into space; curved mirrors can be much larger than lenses so it can collect much more light to see distant objects

Page 91 Waves

1 Energy; matter; oscillations

2 a B **b** E or A

3 Frequency of wave = 15 waves/60 seconds = 0.25 Hz; speed = wavelength × frequency = 8 × 0.25 = 4; m/s

4 a Reaction-time errors; because it is a very short time

b Average time is 0.578 s; speed = distance/time so 200/0.578; = 346 m/s

5 Wavelength = speed/frequency; 1500/500 = 3 m

6 Longitudinal – sound wave; transfer energy; transverse – radio wave; vibrations at right angles to the direction of travel; light waves; transfer energy

7 a A wave caused by earthquakes or explosions; that travels through the Earth

b It detects the vibrations of the earth caused by seismic waves; the movement of the earth is detected relative to a heavy pendulum

Page 92 The electromagnetic spectrum

1 a Transverse; vacuum

b Gamma rays; ultraviolet; infrared; radio waves

2 a Radio waves

b Gamma rays

c Any two from: apply suntan lotion; wear sunglasses; cover our skin with clothes

3 The spectrum of white light is all the colours light is made up of / the colours of the rainbow (red, orange, yellow, green, blue, indigo, violet); you can produce it by shining a ray of light through a glass prism

4 a Herschel found that the temperature was highest; beyond the red end of the spectrum where there was no visible colour; this meant that there were waves with wavelengths longer than visible light

b Ritter found that the reaction was fastest; beyond the area of violet light in the spectrum; this meant that there were waves with shorter wavelengths beyond the violet end of the visible light spectrum

5 a Wavelength = speed / frequency; $3 \times 10^8 / 60 \times 10^6 = 5$ m

b Radio wave

Page 93 Uses of EM waves

1 Microwaves = mobile-phone communication; X-rays = airport-security scanner; infrared = TV remote control; ultraviolet = water sterilisation

2 The microwaves travel in a straight line; from a transmitter/aerial to a satellite; the satellite retransmits the signal to a receiver/aerial in the USA

Answers

3 Any three from: take an infrared photograph of some homes; the more infrared emitted, the lighter the image will be; the better the insulation of the home, the less infrared emitted (or vice versa); the better-insulated homes will be darker on the infrared photograph

4 a i Ray of light refracted; away from the normal

ii Ray of light reflected; angle *l* = angle *r*

b Diagram should show reflections off inside of fibre; emerging from far end; reflected angles should be roughly equal to incident angles

Page 94 Gamma rays, X-rays, ionising radiation

1 X-rays can be used to investigate broken bones; the benefits outweigh the risks

2 Similarity – they are both highly ionising / both can travel through solids; difference – gamma rays come from the nucleus of radioactive atoms while X-rays come from metals bombarded with electrons

3 High-energy electrons from a heated cathode; are accelerated through a very high-voltage tungsten anode; electrons collide with the tungsten atoms and emit X-rays

4 Alpha; beta; gamma

5 a Michael is correct; radioactivity is not affected by physical conditions such as temperature; it only depends on the amount of radioactive material in the sample

b An alpha/beta particle enters the GM tube; and ionises the gas inside the tube; allowing a pulse of electric current to flow; each count represents a single particle

Page 95 The Universe

1 Comet, Moon, Mercury, Jupiter, Sun, Milky Way (3 marks if all correct, 1 mark deducted for each object incorrectly placed)

2 a 11.20/0.38 = 29.5 times larger

b 5.20/1.52 = 3.4 times further away

c 1.00/0.27 = 3.7 times larger

3 a Time in seconds = 8 × 60 = 480; distance = speed × time = $3 \times 10^8 \times 480$; = 144×10^9 m

b Light from distant stars takes several years to reach Earth; so we see the stars as they were when the light left them

4 Stars and galaxies give out different types of EM radiation; some objects do not emit any visible light so can only be seen using other parts of the spectrum

5 a Any two from: content of soil; content of rocks; detect signs of water; detect microorganisms such as bacteria

b SETI is a project that is searching for intelligent life forms elsewhere in the Universe; a radio signal is sent into space; in the hope that another form of life will try to communicate

Page 96 Analysing light

1 a A device to split up light into its component colours

b Use an old CD or DVD as a reflector with a box with a slit in it; let light enter the box via the slit; to reflect off the disc to see the spectrum

2 a Different stars are different temperatures; so they emit different-coloured light

b The chemical composition of the star; the temperature of the star

3 Oxygen and ozone absorb mostly ultraviolet rays; UV rays cause skin cancer; so if there are holes in the ozone, more UV will reach Earth

4 a 3 mm

b It has reduced

c The pitch would be higher

5 Red-shift is when a light source moves away from you; the wavelength appears to lengthen; because red light has a longer wavelength than blue light

Page 97 The life of stars

1 E, C, D, B (3 marks for all correct, 1 mark deducted for each letter in the incorrect position)

2 a The force of gravity is balanced; by outward pressure from the nuclear reactions

b The force of gravity will become larger than the outward force; and the Sun will shrink to become a white dwarf

3 a They are hotter than average-sized stars like the Sun, so they emit more blue light

b Any four from: it will expand to become a super red giant; some heavier elements will start to be produced in the core; nuclear fusion reactions eventually stop; the star collapses due to gravity; an explosion called a supernova occurs; leaving a neutron star/black hole

4 White dwarf, main sequence star, red giant, supernova (3 marks for all correct, 1 mark deducted for each object in the incorrect position)

5 Black holes have such an enormous force of gravity; that no EM radiation is emitted from them so they cannot be observed

Page 98 Theories of the Universe

1 Big Bang theory to ideas 2, 3 and 4; Steady State theory to ideas 1, 2 and 5 (1 mark for each correct connection)

2 a The red-shift; of light from stars

b The left-over radiation; from the Big Bang

c The Steady State theory of the Universe cannot account for the existence of cosmic background radiation; so the only theory that supports it is the Big Bang theory

3 a The detection of waves; of a different wavelength to those you are expecting to record

b To get rid of all known sources; so if the background radiation was still detected, it must come from space

c Because the radiation from the Big Bang; would have been emitted equally in all directions

d The original radiation from the Big Bang would have been in the gamma spectrum; because of the Doppler effect / expansion of the Universe; the wavelength would have increased into the microwave region

Page 99 Ultrasound and infrasound

1 a 50 Hz; 2000 Hz

b Higher

c Infrasound is sound with a frequency too low for humans to hear / below 20 Hz; ultrasound is sound with a frequency too high for humans to hear / above 2000 Hz

2 a Pulses of ultrasound are passed through the womb; they reflect from different surfaces; which allows an image to built up of the unborn child

b Ultrasound does not harm the baby inside the womb; it allows doctors to check for healthy development

3 Man-made – nuclear explosion / drilling for oil; natural – earthquake / volcanic eruption / animal movement / meteorite strikes

4 a 6 milliseconds / 0.006 seconds

Answers

b Depth = speed × time/2; 1500 × 0.006/2; = 4.5 m

c The reflected pulse will move; to the right

d Emit the pulses less frequently

Page 100 Earthquakes and seismic waves

1 a Crust; mantle; outer core; inner core

b Semi-solid rock beneath Earth's crust

2 a Tectonic plates

b Tectonic plates move; due to convection currents in the mantle; earthquakes occur at plate boundaries when the plates suddenly slip

c The coastlines of Africa and South America look like they fit together; similar rock types have been found in Africa and South America; fossils of the same species have been found on both continents

3 a P waves

b P waves

c Both P and S waves

4 Scientists cannot measure the stress/pressure of the rocks; the fault lines are too deep below the ground

5 a A is the P wave and B is the S wave

6 a The P waves are refracted through the core; so there is a region that no P waves can reach

b Waves should start at the epicentre; some should be shown stopping at the core; others should be shown as curved paths reflected off the outer core and stopping at the crust

Page 101 Electrical circuits

1 a

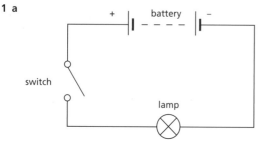

b Any three from: they carry energy from the cell to the bulb; the charged particles in this circuit are electrons; the electrons/charged particles are given energy in the battery; the electrons/charged particles flow all the way round the circuit; at the bulb the energy carried by the electrons/charged particles is converted to heat and light

2 X = 0.3 A; Y = 0.3 A; Z = 0.6 A

3 Bulb B

4 Current; voltage; voltage; voltage; voltage (3 marks for all correct, 1 mark deducted for each incorrect answer)

5 Potential difference is the scientific term for voltage/ measured in volts; it is a measure of how much energy is given to/converted from the charged particles; one volt means that one joule of energy is given to/converted for each unit of charge

Page 102 Electrical power

1 a Kettle

b Iron

c Iron

d

Appliance	Energy used in joules	Energy used in kW h
Kettle	1 200 000 J	0.33 kW h
Iron	2 160 000 J	0.6 kW h
Vacuum cleaner	1 200 000 J	33 kW h

e The watt is too small a unit / the number of watts used by domestic appliances is too large a number

f Kettle – current = 2000/230 = 8.7 A; 13 A fuse; iron – current = 1200/230 = 5.2 A; 13 A fuse; vacuum cleaner – current = 1000/230 = 4.3 A; 5 A fuse

2 a Any two from: less pollution; lowers energy bills; longer-lasting; saves energy resources

b Initial cost of bulbs has to be paid; but the savings on energy bills can be used to offset the cost; payback time is the amount of time taken for the total savings to add up to the initial cost

3 a

Device	Initial cost (£)	Annual saving (£)	Payback time (years)
Double glazing	7000	350	20
Loft insulation	450	75	6
Draught excluders	40	5	8
Cavity-wall insulation	550	110	5

b Double glazing; it increases sound-proofing / increases the value of the property

Page 103 Energy resources

1 E, D, B, C (3 marks for all correct, 1 mark deducted for each sentence in the wrong place)

2 a Carbon dioxide; sulfur dioxide

b They contribute to the greenhouse effect; cause pollution / acid rain

c Any two from: radioactive waste is produced; the consequences of an accident can be disastrous; the start-up time is very long; building/decommissioning a nuclear power station is very expensive

3 Infrared radiation emitted from the surface of the Earth; is absorbed by gases in the atmosphere, making the temperature rise; causing it to re-radiate infrared radiation

4 Renewable – wind, biomass, hydroelectric, solar, tidal; non-renewable – coal, oil, nuclear (half a mark for each correct answer)

5 a Wind turns the turbine blades; which are connected to a generator

b Any two from: they cause noise pollution; they are unsightly in the countryside; they need a large area of land for several turbines; they only produce electricity when the wind is blowing

6 a 2.3 MW (a range of 2.2–2.4 MW acceptable)

b 15 m/s

c No electricity is produced at speeds lower than about 4 m/s; if the wind is too strong, the turbine has to be switched off to avoid damage

Answers

Page 104 Generating and transmitting electricity

1 a Current

   b Moves to the right

   c Stays still

2 a The output voltage would be higher

   b The output voltage would be higher

   c Graph should show a negative sine wave; with the same amplitude

3 a Step-down; step-up; step-down; neither

   b V_1, V_2 V_3, V_4

4 a To reduce energy loss in the cables; because lower currents are needed

   b Cables are either buried underground; or suspended high up using pylons

5 The output from a battery is direct current; which will create a constant magnetic field; for a voltage to be induced on a secondary coil; there must be a varying magnetic field

Page 105 Energy and efficiency

1 Roller coaster = gravitational potential energy; rubber band = elastic potential energy; battery = chemical energy; cup of tea = heat energy

2

Device	Energy input	Useful energy output	Wasted energy output
Electric fan	Electricity	**Kinetic**	Heat and sound
Television	**Electricity**	Light and sound	**Heat**
Catapult	**Elastic potential**	Kinetic	Heat
Gas ring on a cooker	**Chemical**	**Heat**	Light and sound

3 Total energy remains constant; energy can only be transferred; from one type to another / from one object to another

4 a Chemical; heat; kinetic; electrical

   b Efficiency = 144/360; = 0.4/40%

   c It is wasted as heat; in the atmosphere

5 Sankey diagram showing chemical energy in coal at start, with four arrows detailing: heat energy wasted from burning coal; heat energy wasted in turbine; heat energy wasted in generator; useful electrical energy from generator

Page 106 Radiated and absorbed energy

1 a ii 200 ml of water at 40 °C

   b i A cup of tea in a black cup

   c i A black water bottle

2 a 20 °C

   b 20 °C

   c Some of the energy was used to heat up the ice cube; some energy was used to heat up the water; some energy was used to heat up the air

3 a Correctly plotted all points on a suitably scaled axis (3 marks); line of best fit drawn (1 mark)

   b Any two from: use the same amount of water in each beaker; make sure the beakers are the same size; use the same starting temperature; make sure the beakers are at the same room temperature

   c Cooling curve; below the plotted line

   d The black surface will emit more; infrared/heat radiation; so the water will cool more quickly

Page 107 P1 Extended response question

0 marks
Insufficient or irrelevant science. Answer not worthy of credit.

1–2 marks
Answer may be simplistic. There may be limited use of specialist terms. Errors of grammar, punctuation and spelling prevent communication of the science.

3–4 marks
For the most part the information is relevant and presented in a structured and coherent format. Specialist terms are used for the most part appropriately. There are occasional errors in grammar, punctuation and spelling.

5–6 marks
All information in answer is relevant, clear, organised and presented in a structured and coherent format. Specialist terms are used appropriately. Few, if any, errors in grammar, punctuation and spelling.

P2 Physics for your future

Page 108 Electrostatics

1 a Protons; neutrons

   b The protons have a positive charge; the neutrons have no charge

2 a There is friction between the jumper and Brendan's body

   b Electrons

   c All the hair has the same charge; and like charges repel each other

3 a The charged balloon repels some of the electrons away from the surface of the wall; leaving the wall with a positive charge; the opposite charges attract

   b Induction / induced charges

4 a They are the same

   b Ion

   c It loses electrons

5 The charged rod repels/attracts electrons down/up the metal rod of the gold-leaf electroscope away from/towards the cap; the rod and the gold leaf are charged with a negative/positive charge; they repel one another, causing the leaf to rise

Page 109 Uses and dangers of electrostatics

1 Less insecticide can be used; the plant gains an even coating of insecticide; every part of the plant's leaves attracts the insecticide, even the underside

2 a Field lines

   b A stronger field

3 He will experience a small electric shock; because the electrons on the surface will jump the tiny distance of air; and travel through him to the earth

4 a The fuel rubs against the pipe, causing static; the charges can build up and create spark, igniting the fuel

   b The cable will channel the charges safely to earth; where they are dispersed

Answers

5 a Embedded in the ground

   b A positive charge

Page 110 Current, voltage and resistance

1 The electrons travel in one direction only

2 a 2.0 A

   b 2.0 A

3 a Parallel

   b Series

4 a In parallel

   b An ammeter

   c In series

5 7.0 V

6 a 0.15 J

   b 30 V

Page 111 Lamps, resistors and diodes

1 a Potential difference; current

   b
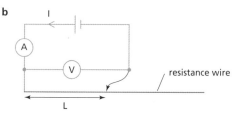

   c The resistance of the wire is directly proportional to its length; when the length doubles, the resistance doubles

2 It is constant

3 a As the current increases the filament gets hotter; which increases resistance

   b The graph would curve; with a decreasing gradient

4 The diode only conducts in one direction; it has an infinite resistance when the current is zero; the diode has low resistance when it conducts

Page 112 Heating effects, LDRs and thermistors

1 The current in the lamp; causes it to heat up; excessive heat could cause a fire

2 a Current (A) = power (W) / potential difference (V); = 4400 / 220; = 20 A

   b Energy (J) = power (W) × time (s); = 4400 × 300; = 1 320 000 J

3 The heating is a result of collisions between electrons; and the ions in the lattice; the collisions cause increased vibrations (around fixed positions) of the ions; which is what we mean by heat energy

4 The resistance of a thermistor decreases as its temperature increases; greater temperature means less resistance

5 a Thermistor

   b LDR

   c Thermistor

6 In thermistors an increase in temperature frees up more electrons from the atoms; and this causes the resistance to decrease

Page 113 Scalar and vector quantities

1 A scalar quantity only has magnitude; a vector quantity has magnitude and direction

2 a 3400 m

   b 1700 m

   c 0 m

3 1.5; m/s

4 a 0.5 m/s^2

   b 36 m/s

   c Its direction is changing; therefore its velocity is changing so it is accelerating

5 a 14 s

   b Yes, the plane will be able to land; the distance required at an average speed of 17.5 m/s is 245 m, which is shorter than the runway at 400 m

Page 114 Distance–time and velocity–time graphs

1 a Speed

   b Its speed is decreasing

2 1.6 / 0.4; = 4.0 m/s^2

3 Constant velocity

4 a It travels with a constant velocity of 30 m/s for 5 seconds; then it decelerates to 0 m/s in a further 3 seconds

   b 10; m/s^2

   c 150 m

   d 45 m

Page 115 Understanding forces

1 a 500 N; forwards

   b Zero

2 It will either remain stationary; or will continue moving with a constant velocity

3 a Newton's third law

   b They are equal; and opposite

4 a The gravitational force; of the Earth on the Sun

   b The gravitational force; of you pulling back on the Earth

Page 116 Force, mass and acceleration

1 Driving force; mass

2 a A motion sensor

   b The mass of the trolley

   c It will be proportional

3 120 – 30 = 90 N so 90 / 90; = 1 m/s^2

4 a 800 N

   b 60 N

5 a Air resistance / drag

   b Upwards

   c He will slow down

   d As he slows down the force of air resistance/drag will reduce; until it balances his weight; the resultant force will then be zero and he will continue at a constant speed

Answers

Page 117 Stopping distance

1 a Reaction time

  b 6 m

2 Thinking distance; braking distance

3 The mass of the car increases; the speed of the car increases; there is reduced friction between the tyres and the road because of a wet or icy surface

4 a When walking / braking

  b Between moving parts of a car; because it wastes energy

5 a 10 m

  b 20 m

  c 15 m

  d 30 m

  e 15 m

Page 118 Momentum

1 a 2400 kg m/s

  b It has both magnitude and direction

2 a 180 kg m/s

  b Zero

  c $(45.0 \times 2.0) + (6.0 \times 0) = (51 \times v) / 51 \times v = 1800$ kg m/s; $v = 1800 / 51$; $= 3.5$ m/s

3 0.5 m/s; in the direction of the 3.0 m/s carriage

4 a It increases the time it takes for the car (and its passengers) to stop; reducing the average force of the impact

  b Safety belts; air bags

5 40 N $\times$ 0.4 s; $= 16$ kg m/s

Page 119 Work, energy and power

1 a The work done when a force of 1 newton moves through a distance of 1 metre; in the direction of the force

  b 100 J

  c 240 J

2 a 4000 J

  b Friction causes energy to be wasted as heat

  c 100 W

3 a 2400 J

  b 60 W

4 20 000 W

Page 120 KE, GPE and conservation of energy

1 a Joules (J)

  b Kilograms (kg)

  c Metres per second (m/s)

2 a 37.5 J

  b Increase the mass; increase the velocity

3 100 000 J

4 a 5400 J

  b It is transferred to kinetic energy

  c 19 m/s

  d It has been transferred to heat; and sound

  e Friction (between the roller blades and the ground)

5 50 m/s

6 It increases by 4; for a given car, the braking force and mass are constants; therefore braking distance is directly proportional to velocity squared

Page 121 Atomic nuclei and radioactivity

1 a Protons; neutrons

  b It has the same number of each

  c Ion

  d Negative

  e Rubbing insulators together

2 a The element's chemical symbol

  b Mass number / nucleon number

  c Atomic number / proton number

  d By working out the difference between A and Z

3 A nuclei of an element with the same number of protons; but a different number of neutrons; it has the same chemical properties; because it has the same number of electrons

4 a Beta particle

  b Alpha particle

  c Gamma ray

5 a 4

  b Stopped by paper, skin or about 6 cm of air

  c −1

  d Very weak

6 Two protons; and two neutrons are removed from the nucleus

Page 122 Nuclear fission

1 a Fusion

  b Fission

  c Neutron

2 Chain reaction

3 a Uranium; plutonium

  b Because nuclear waste can remain radioactive for thousands of years; so it is difficult to dispose of safely

4 a Removes thermal energy produced in the core

  b Slows down the fast-moving neutrons; giving them a greater chance of reacting with other uranium nuclei

  c Absorb the neutrons; to control the chain reaction

5 The minimum mass of a fissile material; required to sustain a chain reaction

Page 123 Fusion on the Earth

1 a Deuterium

  b Tritium

  c Helium

  d Neutron

2 a Nuclear fusion carried out at room temperature

  b Pons and Fleischmann failed to produce sufficient details of their experiment; so their experiment could not be reproduced and therefore validated by other scientists

3 a Uncharged neutrons

  b Because the positively charged hydrogen nuclei repel each other; so they must move rapidly to overcome the electrostatic forces

Answers

c By strong electromagnetic fields; produced by electromagnets

d Because there are such high temperatures and pressures in stars

Page 124 Background radiation

1 Geiger counter

2 a Granite

b Radon

3 a Any three from: nuclear power industry; nuclear weapons tests; air travel; medical; work-related

b Cosmic rays; rocks; food

c Natural sources

4 a Alpha radiation

b Alpha radiation cannot pass through the skin, but radon gas can be inhaled; so the alpha radiation enters the lungs; where it can cause cancer

Page 125 Uses of radioactivity

1 Any two from: irradiating food; detecting cracks in metal; detecting leaks in underground pipes; checking water quality

2 a Beta radiation

b It would get weaker

c It would get stronger

3 It is the most ionising radiation; so it creates more ions in the air than any other radiation; which makes the smoke alarm most effective

4 Equipment can be sterilised when sealed in a package

5 ii, iii, i

Page 126 Activity and half-life

1 a It cannot be predicted

b It is not affected by external conditions

2 a Bequerel (Bq)

b It doubles

c It is inversely proportional

3 a 200 Bq

b 50 Bq

4 Any three from: point the source away from other people; use special tools or gloves to handle the source; only remove the source from its lead-lined container when you need it; wash your hands after using the source

5 Nuclear power stations produce waste that remains radioactive for thousands of years; so disposing of it safely is a problem

6 It can be stored deep underground; or in special tunnels made under mountains.

Page 127 P2 Extended response question

0 marks
Insufficient or irrelevant science. Answer not worthy of credit.

1–2 marks
Answer may be simplistic. There may be limited use of specialist terms. Errors of grammar, punctuation and spelling prevent communication of the science.

3–4 marks
For the most part the information is relevant and presented in a structured and coherent format. Specialist terms are used for the most part appropriately. There are occasional errors in grammar, punctuation and spelling.

5–6 marks
All information in answer is relevant, clear, organised and presented in a structured and coherent format. Specialist terms are used appropriately. Few, if any, errors in grammar, punctuation and spelling.

P3 Applications of physics

Page 128 Intensity of radiation

1 Light waves – non-ionising; ultrasound – non-ionising; ultraviolet – ionising; microwaves – non-ionising

2 a So the only light detected by the light meter has come from the torch

b Sensible scales; 5 points plotted correctly; smooth curve drawn

c As the distance increases the light intensity decreases; because the light is spread out over a larger area as it gets further from the source

d Another smooth similarly shaped curve drawn on the graph grid lower than the original one

3 a $I = P/4\pi r^2$; $4\pi r^2 = P/I = 600/0.5 = 1200$; $r^2 = 1200/4\pi = 95.5$; $r = 9.8$ m

b There are likely to be objects in the way; which will absorb some of the radiation.

Page 129 Properties of lenses

1 a Diverging

b Converging

c Diverging

2 a $1/0.15 = -6.7$ D (2)

b $1/0.12 = 8.3$ D (2)

3 a Combined power = 6; $f = 1/6 = 0.17$ m

b Power = $1/0.083 = 12$ D; Lenses A and C

c Power = $1/-0.5 = -2$ D; Lenses C and E

Page 130 Lens equation

1 a Move the screen backwards and forwards until the image is in focus; use a metre rule to measure the distance from the lens to the screen

b Diminished; real

c The image would be larger and further from the lens

2 One ray drawn horizontally from top of object to lens, which refracts through focal point; one ray drawn from top of object through centre of lens; all rays drawn with ruler; image drawn where two lines cross

3 a $\frac{1}{f} = \frac{1}{0.04} + \frac{1}{2}$; $\frac{1}{f} = 25 + 0.5 = 25.5$; $v = \frac{1}{25.5} = 0.039$ m

b $\frac{1}{3.9} = \frac{1}{3} + \frac{1}{v}$; $\frac{1}{v} = \frac{1}{3.9} - \frac{1}{3} = 0.255 - 0.333 = -0.078$; $v = -\frac{1}{0.078} = -1.13$ m

Answers

Page 131 The eye

1 a **i)** B and C; **ii)** E.

 b It changes the shape of the lens; so that the lens can focus at different distances

2 Any **3** of: The near point is the closest distance the eye can focus; for a normal human eye it is about 25 cm; the far point is the furthest distance the eye can focus; for a normal human eye it is at infinity

3 a Short-sighted

 b Diverging lens

 c Any **2** of: glasses often steam up; people can't see contact lenses; better all round vision – able to see from corner of eye

4 1 advantage, eg: permanently corrects vision; **1** disadvantage, eg: expensive treatment; risk of infection; might be painful

Page 132 Total internal reflection and endoscopes

1 a Ray of light refracted away from the normal

 b Ray of light totally internally reflected with angle i = angle r (by eye)

2 a Ray of light reflected several times at boundary; straight lines and approximate angles of reflection correct (by eye)

 b If the fibre is coiled up tightly the angle the light makes with the boundary will change; if it coils too much the angle will be smaller than the critical angle; so some light will refract out of the fibre

 c Any **2** of: An endoscope is a medical instrument; used for seeing inside cavities of the body; used for minor surgical procedures; light is transmitted along optical fibres inside the body; and an image is transmitted back along different optical fibres

3 a Refractive index for water = 1.32; refractive index for Perspex = 1.47; critical angle for glass = 41.8°; critical angle for diamond = 25.2°

 b i difference in speed of light will be smaller; **ii** straight line drawn on diagram showing less refraction than before; **iii** the critical angle for glass–water boundary will be larger than for glass–air boundary; because the light is not refracted so much/ratio of refractive indices is lower/ it will take a larger angle of incidence for the light to be refracted at 90°

Page 133 Medical uses of ultrasound

1 Very high frequency sound waves; which humans cannot hear/over 20 kHz

2 X-rays are ionising radiation and ultrasound is not; X-rays can harm the cells inside the growing foetus

3 a Pulses are used so the exact time to travel there and back can be recorded

 b Difference in time between top and bottom of head = 0.00028 s; total distance = 1500 × 0.00028 = 0.42 m; distance between top and bottom of head = 0.21 m

4 1 diagnostic use, e.g.: measuring the depth of the eye; monitoring cyst/tumours etc; **1** use for treatment described, e.g.: removing kidney/bladder/gall stones; treating damaged muscles (1 mark for diagnostic use, 1 mark for treatment use)

Page 134 Producing X-rays

1 a So the electrons don't collide with air molecules; and slow down/stop

 b To heat the cathode; so it can emit electrons

 c To accelerate the electrons; towards the metal target

 d The anode gets very hot; the metal may melt

 e Current is a flow of charged particles; electrons are moving so there is a current

 f Increasing the temperature of the cathode

 g Increasing the potential difference between the anode and cathode

2 a $KE = e\,V = (1.6 \times 10^{-19}) \times 65000; = 1.04 \times 10^{-14}$ J

 b $\frac{1}{2}mv^2 = KE = 1.04 \times 10^{-14}; v^2 = 2 \times (1.04 \times 10^{-14})/9.1 \times 10^{-31}; v = \sqrt{(2.29 \times 10^{16})} = 1.5 \times 10^8$ m/s

 c $I = N\,q$ so $N = I/q; = 1.5 \times 10^{-3}/1.6 \times 10^{-19} = 9.4 \times 10^{15}$

Page 135 Medical uses of X-rays

1 a Any **2** of: CAT scans can show any damage to soft tissue; conventional X-rays only show damage to bones; CAT scans can produce a 3-dimensional image

 b X-rays are ionising radiation; X-rays can cause damage to cells

2 Any **4** of: patient has a drink made of barium sulphate; barium absorbs X-rays; so if X-rays are fired towards the intestine; the location of the barium is easily seen; the barium will build up at a blockage

3 a 50, 22.2; 12.5, 8

 b Lead aprons will absorb most of the X-rays; radiographers are exposed to a lot of X-rays, so need to be protected

 c Any **4** of: X-rays do cause damage to cells; so large exposure to X-rays can cause cancer; usually X-rays are focused on the cancer cells; so other healthy cells do not receive a high dose of X-rays; some damage to healthy cells may occur; it is a case of deciding whether the benefits of treatment outweigh the risks; it may be correct to use treatment in some cases, but not others

Page 136 Heart action and ECG

1 a 72 bpm

 b 72/60; = 1.2 Hz

2

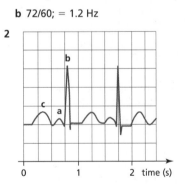

3 a Electrodes are attached to his body; using a conducting jelly; the electrodes are connected to a monitor which displays the electrical activity

 b There will be an uneven shape to the ECG curve; the peaks will be different distances apart

 c Drugs to regulate heartbeat; or a pacemaker

Page 137 Pulse oximetry

1 Pulse rate and amount of oxygen in the blood

2 a Protein molecule; inside red blood cells; which carries oxygen around the body

 b The amount of light absorbed by the haemoglobin depends on the amount of oxygen present; the more oxygen carried, the lower the transmitted light intensity

Answers

c It monitors a patient continuously; giving a record of the pulse rate and/or the amount of oxygen in the blood; so medical professionals can identify any problems

d There are two different forms of haemoglobin; oxyhaemoglobin absorbs infrared radiation; deoxygenated haemoglobin absorbs visible light; with the reading for absorption of both wavelengths an accurate percentage of oxygenated haemoglobin can be calculated

Page 138 Ionising radiations

1 a A particle inside the nucleus; consists of protons and neutrons

  b A version of an element; with the same number of protons; and a different number of neutrons in the nucleus

2 a 8 protons; 8 neutrons

  b 26 protons; 28 neutrons

  c 92 protons; 144 neutrons

3 a Alpha

  b Beta-plus; alpha

  c Neutron; gamma

  d Beta-plus

  e Beta-minus

4 Any 2 of: A dosimeter is a radiation badge; which consists of a layer of film; which shows exposure to ionising radiation

  And any 2 of: Used to monitor the level of radiation exposure; a record over time can be stored; so workers do not exceed annual/monthly/weekly recommended dose

Page 139 Radioactive decays

1 a $^{216}_{84}$Po

  b $^{226}_{88}$Ra

  c $^{234}_{90}$Th

  (2 marks each: 1 mark for the mass number and 1 mark for the atomic number plus corresponding symbol)

2 a A neutron within the nucleus; changes into a proton and an electron; the proton remains in the nucleus and the electron is emitted

  b $^{129}_{54}$Xe (1 mark for the mass number and 1 mark for atomic number plus corresponding symbol)

  c To remove surplus energy from the nucleus

3 a $^{241}_{95}$Am $\longrightarrow$ $^{237}_{93}$Np $+ ^{4}_{2}$He

  b $^{40}_{19}$K $\longrightarrow$ $^{40}_{18}$Ar $+ ^{0}_{+1}$e

  c $^{137}_{55}$Cs $\longrightarrow$ $^{137}_{56}$Ba $+ ^{0}_{-1}$e

  (3 marks each: 1 mark for the mass number of daughter; 1 mark for the atomic number plus symbol; 1 mark for the correct particle)

Page 140 Stability of nuclei

1 a Stable nuclei lie on a gentle upward sloping curve; with usually more neutrons than protons; there is a strong nuclear force between all the nucleons; which overcomes the electrostatic repulsion between the protons

  b Point B marked just above the lower part of the curve

  c Point C marked just below the lower part of the curve

  d Point D marked on or just below the top end of the curve

2 A neutron-poor isotope has too many protons to be stable; one of its protons changes into a positron and a neutron; the positron is emitted as beta-plus and the neutron remains in the nucleus

3 A particle with very low mass; and no charge; which is emitted alongside a beta-plus particle.

Page 141 Quarks

1 a Particles which cannot be subdivided; into smaller particles

  b Quarks (and leptons)

2 a A neutron has 1 up quark and 2 down quarks; a proton has 1 down quark and 2 up quarks

  b Up quark has a charge of $+^{2}/_{3}$e; down quark has a charge of $-^{1}/_{3}$e; $(2 \times ^{2}/_{3})$e $+ (1 \times -^{1}/_{3})$e $= (^{4}/_{3} - ^{1}/_{3})$e $= +1$ e

3 Beta minus decay: within one neutron; a down quark changes into an up quark and an electron

  Beta-plus decay: within one proton; an up quark changes into a down quark and a positron

Page 142 Dangers of ionising radiations

1 a Any 3 of: Skin burns; nausea; hair loss; destruction of bone marrow; changes to genetic material; sterility; cancer

  b Any 2 of: Keep a good distance away from the radioactive source; use shielding, e.g. lead; keep exposure time to a minimum

  c Cells which have been damaged by radiation; repair themselves abnormally; so they malfunction or reproduce at an uncontrolled rate

  d Type of radiation; the dose received; the part of the body receiving the radiation

2 a The effective biological damage to human tissue; measured in Sieverts

  b Any 4 relevant points explained clearly such as: It is relatively safe for the patient; because it is only a quarter of the annual maximum dose; the benefits from the scan will outweigh the risks; the radiographer will administer several CAT scans each day; so will need to wear protective clothing/be shielded from radiation; the dose received outside the machine will be less than 5 mSv

Page 143 Treatment of tumours

1 a Treatment to shrink or slow down growth of a tumour; does not cure the cancer

  b Any 2 of: tiredness; sickness; feeling sore

2 a Intense X-rays; directed at specific area of body; intensity and energy controlled

  b Radioactive substance injected; radioactive metal implant or wires close to site of cancer; radioactive capsule or drink imbibed

3 Neutrons have a large mass; and can have very high speed; so have an enormous amount of (kinetic) energy; so can knock electrons off atoms in a collision

Page 144 Diagnosis using radioactive substances

1 a A radioactive isotope; of an element used by the body in some way; which is swallowed or injected into the body

  b Iodine-131; not alpha emitter – alpha is too ionising inside the body; suitable half life – 6 s and 83 s are too short

Answers

2 a Any **2** of: activity of brain; spread of cancer through the body; flow of blood through organs

b Ordinary radioactive tracers are elements usually used by the body; tracers used in PET scans are tagged onto chemicals used by the body; radiation detectors detect the radiation emitted by an ordinary radioactive tracer; in PET scans the gamma rays emitted by positron/electron annihilation are detected

Page 145 Particle accelerators and collaboration

1 a Machine which accelerates charged particles; to almost the speed of light; either as a linear accelerates or as a ring (LHC)

b Any **2** of: Collide subatomic particles together; to investigate the structure of particles; to investigate the behaviour of particles; to develop a better understanding of the physical world; to prove the existence of new particles

c i So it does not take up as much space; particles can accelerate for several revolutions; particles can be sent both clockwise and anticlockwise around the ring

ii Circumference of the circle $= \pi d = \pi 9000 = 2.8 \times 10^4$ m; 11 000 revolutions per second $= 3.1 \times 10^8$ m

2 a Any **2** of: Sharing the cost between many nations; no political barriers; wider pool of expert scientists from around the world; sharing ideas between many scientists

b Any **3** salient points explained, e.g.: Benefits of pure research; there may never be any commercial application for the new discoveries; military uses; all results in public domain; corruption

Page 146 Cyclotrons

1

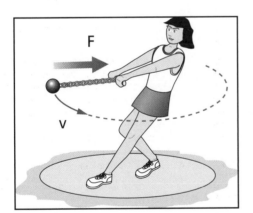

c Centripetal force

2 a Provide a force on the charged particles; at right angles to the motion of the particles; to provide centripetal force

b Particles move in semicircular path in dees; then accelerated by electric field; speed increases so radius of curvature of circular path increases

3 a $^{13}_{7}$N (1 mark for mass number; 1 mark for atomic number plus symbol)

b $^{11}_{6}$C $+ ^{4}_{2}$He (1 mark for mass number; 1 mark for atomic number plus symbol; 1 mark for correct He numbers)

c $^{15}_{8}$O (1 mark for mass number; 1 mark for atomic number plus symbol)

Page 147 Electron–positron annihilation

1 a Converted to energy; in the form of gamma rays

b Two gamma rays; go in opposite directions

c Positron has one positive charge and electron has one negative charge; so overall initial charge is zero; gamma rays emitted have no charge

2 Positrons which are emitted by the radioisotope; interact with electrons in the body; emit gamma rays which are detected; the time delay and position of the detections is used to locate the position

3 Rearrange $E = mc^2$ to give $m = E/c^2$; $m = 3.9 \times 10^{-12}/(3 \times 10^8)^2$ $= 4.3 \times 10^{-29}$ kg; number of particles $= 4.3 \times 10^{-29}/9.11$ $\times 10^{-31} = 48$ particles so 24 pairs

Page 148 Momentum

1 a 10×0.05; $= 0.5$ kg m/s

b 0 kg m/s

c Initial momentum = final momentum; $0.5 = 0.25$ v; v $= 2$ m/s

2 a Kinetic energy is conserved in elastic collisions; but not in inelastic collisions

b Energy before collision $= \frac{1}{2}mv^2$; $= \frac{1}{2}(0.05)(10^2) = 2.5$ J; energy after collision $= \frac{1}{2}(0.25)(2^2) = 0.5$ J; energy difference $= 2$ J

c Lost to environment as sound/heat; some used to deform balls slightly

3 Final momentum of bullet and wood = total mass × final velocity; $= 0.52 \times 1.3 = 0.676$ kg m/s; initial velocity of bullet $= 0.676/0.02 = 33.8$ m/s

Page 149 Matter and temperature

1 Any **3** of: molecules are vibrating about fixed positions when solid; as the temperature rises they vibrate faster; some will break free from bonds; and will move about freely in liquid form

2 Temperature at which molecules would not move at all

3 308 K

4 –97 °C

5 a The particles move faster; and become more spread out

b KE $\propto$ T so $\frac{1}{2}mv^2 \propto$ T; $v^2 \propto$ T or; $\dfrac{v_1^2}{v_2^2} = \dfrac{T_1}{T_2}$; rearrange to give

$v_1^2 = \dfrac{v_2^2\, T_1}{T_2} = (500)^2 \times 313/293$; v1 $= \sqrt{2.67 \times 10^5} = 516$ m/s

Page 150 Investigating gases

1 a Molecules move around fast; colliding with the walls of the cylinder; the force of all the molecules hitting the wall creates a pressure

b F $= 320 \times 10^5 \times 0.16$; $= 5.12 \times 10^6$ N

c $P_1V_1 = P_2V_2$ rearranged to give $P_1 = \dfrac{P_2V_2}{V_1}$;

$P_1 = 320 \times 10^5 \times 2 \times 10^{-3}/3 \times 10^{-3}$; $= 213 \times 10^5$ Pa

d Particles move about, colliding with walls of container; if volume increases the particles won't hit the walls so often; so pressure will decrease

2 a As temperature increases particles will move faster; and collide more frequently; so they will move further apart from each other

b $\dfrac{V_1}{T_1} = \dfrac{V_2}{T_2}$ rearranged to give $V_1 = \dfrac{V_2 T_1}{T_2}$;

$T_1 = 70 + 273 = 343$ K and $T_2 = 25 + 273 = 298$ K;

$V_1 = 7 \times 10^{-6} \times 343 / 298$; $= 8 \times 10^{-6}$ m³

Answers

Page 151 The gas equation

1 a The temperature of the gas; the pressure of the gas

b As the temperature increases the pressure will increase; because as the temperature increases the speed of the particles increases; so they will hit the walls of the container more often

2 a C

b B

c D

3 a $\dfrac{P_1 V_1}{T_1} = \dfrac{P_2 V_2}{T_2}$ rearranged to give $V_1 = \dfrac{P_2 V_2 T_1}{P_1 T_2}$;

$T_1 = 298$ K and $T_2 = 283$ K; $V_1 = (200 \times 10^5 \times 0.02 \times 298)/1 \times 10^5 \times 283$; $= (1.2 \times 10^8 / 2.83 \times 10^7) = 4.2$ m³

b Convert 4.2 m³ to 4200 litres; time = 4200/2.5 = 1680 minutes

Page 152 Extended response question

0 marks
Insufficient or irrelevant science. Answer not worthy of credit.

1–2 marks
Answer may be simplistic. There may be limited use of specialist terms. Errors of grammar, punctuation and spelling prevent communication of the science.

3–4 marks
For the most part the information is relevant and presented in a structured and coherent format. Specialist terms are used for the most part appropriately. There are occasional errors in grammar, punctuation and spelling.

5–6 marks
All information in answer is relevant, clear, organised and presented in a structured and coherent format. Specialist terms are used appropriately. Few, if any, errors in grammar, punctuation and spelling.